Studies in Big Data

Volume 36

Series editor

Janusz Kacprzyk, Polish Academy of Sciences, Warsaw, Poland
e-mail: kacprzyk@ibspan.waw.pl

The series "Studies in Big Data" (SBD) publishes new developments and advances in the various areas of Big Data- quickly and with a high quality. The intent is to cover the theory, research, development, and applications of Big Data, as embedded in the fields of engineering, computer science, physics, economics and life sciences. The books of the series refer to the analysis and understanding of large, complex, and/or distributed data sets generated from recent digital sources coming from sensors or other physical instruments as well as simulations, crowd sourcing, social networks or other internet transactions, such as emails or video click streams and other. The series contains monographs, lecture notes and edited volumes in Big Data spanning the areas of computational intelligence incl. neural networks, evolutionary computation, soft computing, fuzzy systems, as well as artificial intelligence, data mining, modern statistics and Operations research, as well as self-organizing systems. Of particular value to both the contributors and the readership are the short publication timeframe and the world-wide distribution, which enable both wide and rapid dissemination of research output.

More information about this series at http://www.springer.com/series/11970

Joanna Kołodziej · Florin Pop
Ciprian Dobre
Editors

Modeling and Simulation in HPC and Cloud Systems

Springer

Editors
Joanna Kołodziej
Institute of Computer Science
Cracow University of Technology
Cracow
Poland

Ciprian Dobre
Department of Computer Science
University Politehnica of Bucharest
Bucharest
Romania

Florin Pop
Department of Computer Science
University Politehnica of Bucharest
Bucharest
Romania

ISSN 2197-6503 ISSN 2197-6511 (electronic)
Studies in Big Data
ISBN 978-3-319-89258-0 ISBN 978-3-319-73767-6 (eBook)
https://doi.org/10.1007/978-3-319-73767-6

Printed on acid-free paper

This Springer imprint is published by Springer Nature
The registered company is Springer International Publishing AG
The registered company address is: Gewerbestrasse 11, 6330 Cham, Switzerland

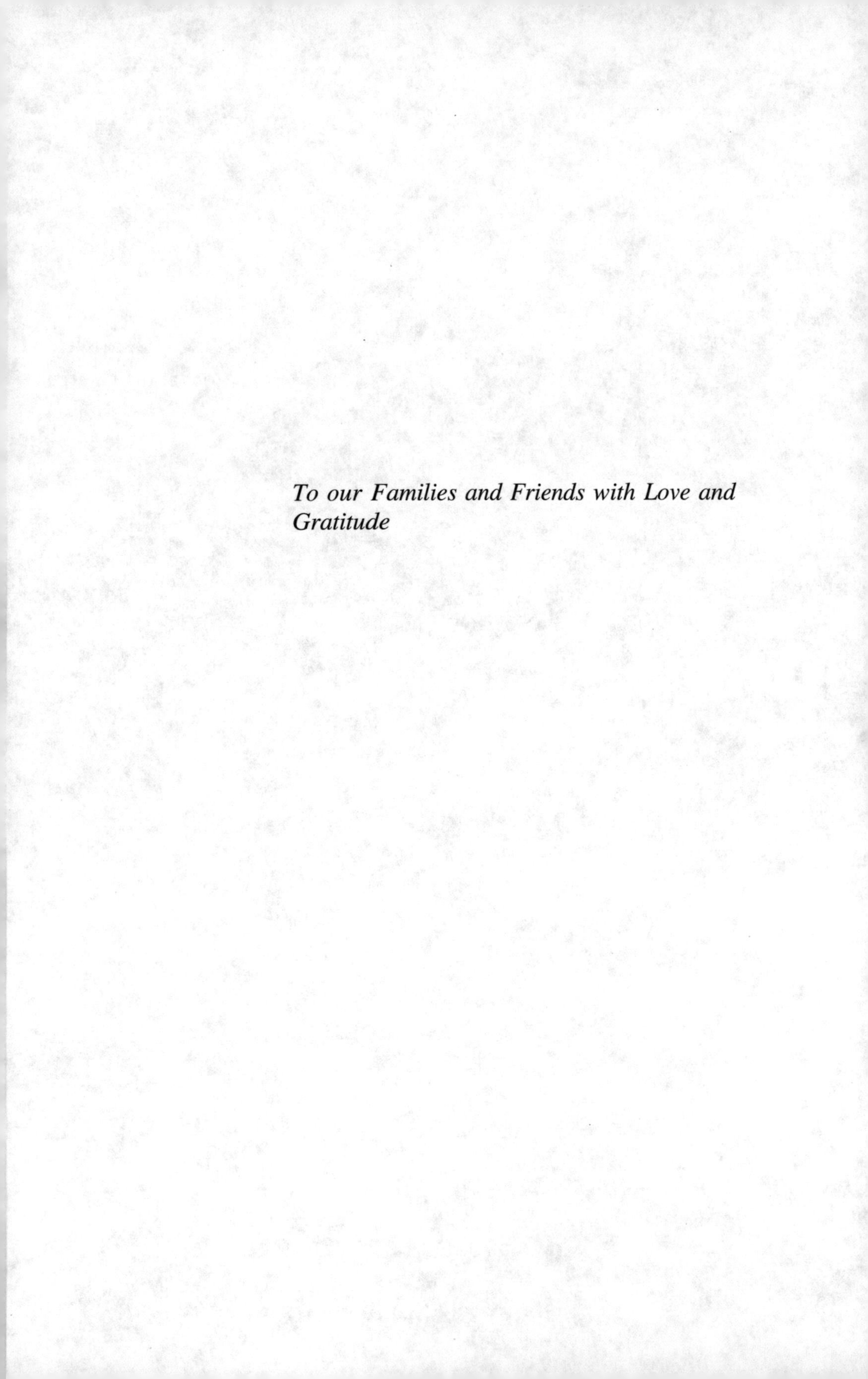

To our Families and Friends with Love and Gratitude

Preface

The latest trend of computing paradigm is to push elastic resources such as computation and storage to the edge of networks by splitting computation between fog nodes (devices, set—top—boxes and datacenters). Mobile applications that utilize Cloud computing technologies have become increasingly popular over the recent years, with examples including data streaming, real-time video processing, IoT data gathering and processing, social networking, gaming, etc. Such applications generally consist of a frontend component running on the mobile device and a backend component running on the Cloud, where the Cloud provides additional data storage and processing capabilities. With this architecture, it is possible to run complex applications on handheld devices that have limited processing power. A mobile device is a computing and engineering marvel, but it has limitations in terms of what it can do. Whether the mobile device is an ultra-book, a tablet, or a smartphone, it will have local storage and native applications, and it will perform most tasks locally just fine. Nonetheless, the Cloud will help the mobile device do the "heavy lifting" when necessary. These models follow the "Smart Anything Everywhere" trend that requires new models for application orchestration and development. These models should be cost-aware and energy-aware in using computation and communication.

Most of the existing Cloud providers (Amazon EC2, Google Cloud Platform, Microsoft Azure etc.) use replication as the fundamental mechanism to provide fault tolerance at data level (Amazon Elastic Block Store, GlusterFS, GFS, HDFS). The motivation for this approach lies in the low cost of the Cloud components: commodity disks, processors, memory, chip sets, etc. As for computing tasks, the main approach is re-execution (Hadoop task and node fault-tolerance model). Tasks are independent: they do not communicate with each other. The reduce part of the Map-Reduce computational model is a task by itself that could be restarted in case of failure. As a consequence, when a task fails, other tasks are not affected. Task re-execution could be accelerated by the use of check-pointing or incremental check-pointing. However, this is not a common situation in Clouds, where storage is mainly used for application data[1]. The advance modelling and performance evaluation are the keys aspects for Cloud Providers.

Middleware infrastructures designed to manage distributed applications typically rely on several layers. Although each previously proposed solution in the literature adds its own layers, there are certain components generally present in all content management systems[2]. The bottom level contains various sensors (physical, virtual or logical), which include all types of sensing devices, from specialized equipment such as air pollution analysers, to smartphones or tablets. The next layer deals with retrieval of raw context data and provides standardized APIs to higher levels. The third layer is responsible for pre-processing functions and is used when raw information is too large or coarse-grained or it needs translation to a specific format. The storage and management layer is at the next level and it handles the way data is stored and how interested entities can request it, using either push- or pull-based access. At top level lies the application layer, which is basically the client. Context models are employed in order to represent gathered information in an efficient and easily-accessible way[3]; they typically include key-value, mark-up schemes, and graphical, object-oriented, logic-based or ontology-based models.

Most applications require content gathering networks built on the top of Cloud systems. The main challenge is that there are already millions of mobile devices in use, which can be utilized at any time. However, this also represents a disadvantage due to the heterogeneity of these devices. Because of the high levels of mobility (as well as energy levels, communication channels or even owner preferences), the accuracy, latency, and confidence of the data that these devices produce can fluctuate heavily at times. Therefore, the main challenge in content gathering networks is identifying the best set of devices that could offer required data, but also configuring them with suitable parameters in order to adjust the results' quality. Cloud systems are used for autonomous computing, a paradigm introduced by IBM[4] that aims at building systems that can manage themselves by automatically adapting to the changing environment conditions. An autonomic system is able to continually react to external factors and update its state according to high-level policies. To achieve self-management, a system has to encapsulate a set of self-* properties, including the main four properties defined by the initial IBM autonomic computing initiative: self-configuration, self-protection, self-optimization, and self-healing. Subsequent research efforts extended this list with new properties conceived to

---

[1] Olofsson, Kristoffer Andreas. "Gathering and organizing content distributed via social media." U.S. Patent 8,990,701, issued March 24, 2015.

[2] Vincent C. Emeakaroha, Kaniz Fatema, Philip Healy, and John P. Morrison. "Contemporary Analysis and Architecture for a Generic Cloud-based Sensor Data Management Platform." Sensors & Transducers 185, no. 2 (2015): 100.

[3] Yaser Jararweh, Mahmoud Al-Ayyoub, Elhadj Benkhelifa, Mladen Vouk, and Andy Rindos. "Software defined Cloud: Survey, system and evaluation." Future Generation Computer Systems (2015).

[4] An architectural blueprint for autonomic computing. Tech. rep., IBM, 2003.

enable or to complement the initial ones, such as self-awareness, self-adjusting, self-anticipating, self-organizing, and self-recovery. A comprehensive reference model, FORMS (FOrmal Reference Model for Self-adaptation) targets highly dynamic and unpredictable behavior in distributed systems[5].

There is a very thin theoretical effort to characterize the resistance of distributed systems to complex faults or perturbing events (failures, cyber-attacks, internal re-organization, or unexpected load) that may happen simultaneously. Recently, theoretical studies investigate algorithmic building blocks for distributed systems that may tolerate both Byzantine faults and transient faults or special kinds of mobile Byzantine faults, which advanced the research in the field. These works concern only two building blocks in distributed systems (agreement and storage) and investigate only simple communication models (synchronous communication). Note that, these studies have been done by members of cHipSet COST Action and due to their recent publication (2016) none of the existing Cloud systems integrate yet these theoretical advances.

Other challenges include: localized analytics, resource limitation, privacy, security and data integrity, aggregate analytics and architecture, presented in detail in[6]. Localized analytics are needed because raw data may not be suitable for direct consumption by the application, so in a way they are the equivalent of the pre-processing layer of the context-aware system architecture. However, it should be noted that pre-processing of raw data should be performed in regard to the context of the requesting node, therefore the challenge in this case is designing the appropriate heuristics to achieve this (e.g. data mediation and context inference).

Cloud computing systems are also built for Big Data Processing. The main research challenges and issues are: (i) the cloud infrastructure is increasingly more heterogeneous (multi-core processors, attached co-processors like GPUs and FPGAs); (ii) the spectrum of cloud services and cloud applications widens (e.g. in the last year Amazon EC2 added 10+ new services, the cloud infrastructure is expected to support an increasing number of data- and CPU-intensive Big Data applications); (iii) Cloud over-provisioning demands high initial costs and leads to a low system utilization (average cloud server utilization is in the 18% to 30% range, but this strategy is not economically sustainable) and (iv) the cloud computing landscape is fragmented (Cloud service providers support different cloud delivery models: Amazon—IaaS, Microsoft—PaaS and Google mostly SaaS)[7].

---

[5] Danny Weyns, Sam Malek, and Jesper Andersson. "FORMS: Unifying reference model for formal specification of distributed self-adaptive systems" (2012). ACM Trans. Auton. Adapt. Syst. 7, 1, Article 8 (May 2012), 61 pages.

[6] Ganti, R. K.; Fan Ye; Hui Lei; "Mobile crowdsensing: current state and future challenges" (November 2011) Communications Magazine, IEEE, vol. 49, no.11, pp. 32–39.

[7] Dan Marinescu, "Self-organization and the Cloud Ecosystem", Invited Talk at The 20th International Conference on Control Systems and Computer Science, May 2015, Bucharest, Romania.

Other new aspects related to Cloud systems were presented in several special issues, published by the authors: Autonomic Computing and Big Data platforms published in Soft Computing Journal[8] being oriented on computer and information advances aiming to develop and optimize advanced system software, networking, and data management components to cope with Big Data processing; Advanced topics on middleware services for heterogeneous distributed computing (Part 1 and Part 2), being oriented on modeling of resource management and middleware systems different computing paradigms like Cluster Computing, Grid Computing, Peer-to-Peer Computing, and Cloud Computing all involving elements of heterogeneity[9,10] published by Elsevier in Future Generation Computer Systems (FGCS). Finally, the book published in Springer, Resource Management for Big Data Platforms: Algorithms, Modelling, and High-Performance Computing Techniques[11] serves as a flagship driver towards advance research in the area of Big Data platforms and applications being oriented on methods, techniques, and performance evaluation.

Modelling and Simulation (MS) in the Big Data era is widely considered the essential tool in science and engineering to substantiate the prediction and analysis of complex systems and natural phenomena. MS offers suitable abstractions to manage the complexity of analyzing Big Data in various scientific and engineering domains. Unfortunately, Big Data problems are not always easily amenable to efficient MS over HPC. Also, MS communities may lack the detailed expertise required to exploit the full potential of HPC solutions, and HPC architects may not be fully aware of specific MS requirements.

This book herewith presents the comprehensive analysis of state-of-the-art, solid backgrounds and novel concepts in the analysis, implementation, and evaluation of next generation HPC and cloud–based models, technologies, simulations and implementations for data intensive applications. The book contains eight chapters, where five are the summary of the tutorials and workshops organized as a part of the Traiing School of the cHiPSet (High-Performance Modelling and Simulation for Big Data Applications http://chipset-cost.eu) Cost Action on "New Trends in Modeling and Simulation in HPC Systems," held in Bucharest (Romania) on September 21–23, 2016. The main goal of that school was to improve the participants' practical skills and knowledge of the novel HPC-driven models and

[8]Pop, Florin, Ciprian Dobre, and Alexandru Costan. "AutoCompBD: Autonomic Computing and Big Data platforms." (2017): 1–3.

[9]Pop, Florin, Xiaomin Zhu, and Laurence T. Yang. "Midhdc: Advanced topics on middleware services for heterogeneous distributed computing. part 1." Future Generation Computer Systems 56 (2016): 734–735.

[10]Pop, Florin, Xiaomin Zhu, and Laurence T. Yang. "MidHDC: Advanced topics on middleware services for heterogeneous distributed computing. Part 2." (2017): 86–89.

[11]Pop, Florin, Joanna Kołodziej, and Beniamino Di Martino, eds. Resource Management for Big Data Platforms: Algorithms, Modelling, and High-Performance Computing Techniques. Springer, 2016.

technologies for Big Data applications. The trainers, who are also the authors of the chapters of this book, explained how to design, construct, and utilize the complex MS tools that capture many of the HPC modeling needs, from scalability to fault tolerance and beyond. The remaining three chapters present the first results of the school: new ideas and novel results of the research on security aspects in clouds, first prototypes of the complex virtual models of data in Big Data streams and data intensive computing framework for Opportunistic Networks. All that work was realized by the trainees in their research teams.

cHiPSet Training School and Research Material

cHiPSet Training School 2016

The training school within COST Action IC1406[12] featured presentations and hands-on practice and demonstration of novel methods, mechanisms, techniques, and technologies in Modelling and Simulation (MS), with a special emphasis on evaluation of HPC Systems. Today MS is widely considered the essential tool in science and engineering to substantiate the prediction and analysis of complex systems and natural phenomena. MS offers suitable abstractions to manage the complexity of analyzing Big Data in various scientific and engineering domains. Unfortunately, Big Data problems are not always easily amenable to efficient MS over HPC. Also, MS communities may lack the detailed expertise required to exploit the full potential of HPC solutions, and HPC architects may not be fully aware of specific MS requirements. Thus, the goal of the training school was to offer to participants, Ph.D. students and Early Stage Researchers, coming from these two worlds, the skills to understand and work with models and concepts coming from HPC, to design accurate modeling and simulation strategies for the evaluation of HPC solutions, to design, construct, and use complex MS tools that capture many of the HPC modeling needs, from scalability to fault tolerance and beyond. At the end, participants learned how to efficiently turn massively large HPC data into valuable information and meaningful knowledge, with the help of covered new trends in MS.

The logical structure of the Training School programme, in terms of reached topics and subjects, allowed participants to understand and grasp concepts related to the terminology, properties and the models used to evaluate HPC systems using modeling and simulation. This was continued with a set of lectures and hands-on exercises, on tools to evaluate using modeling and simulation systems and applications, either related to distributed processing infrastructures, Cloud systems, or various other applications. In the end, the Training School ended with lectures on what does it mean to develop good evaluation methodologies.

---

[12] www.chipset-cost.eu.

TS Research Results

The first five chapters in this part are the etended version of the training material presented at the cHiPSet training school. They contain the background information of the HPC modelling and simulation domain in the Big Data modern era. New trends, models, and challenges are addressed in this part.

In Chapter "Evaluating Distributed Systems and Applications Through Accurate Models and Simulations", Frincu et al. discuss the role of modeling and simulation in the callibration and evaluation of the HPC systems implemented in the distributed environments. The chapter focuses on the theory and hands–on behind some of the most widely used tools for simulating a Cloud environment. The authors analyzed selected existing models for representing both applications and the underlying distributed platform and infrastructure.In simulation part, they used SimGrid simulator. The first author and TS trainer—Dr. Marc Frincu—is a young promising researcher from the Department of Computer Science at UVT working on cutting edge topics related to clouds, smart grids, and Big Data. He did his research also at the University of Southern California working with prof. Viktor Prasanna on smart grids and clouds. At UVT he leads the CER research group focusing on applying cloud computing to areas such as smart grids and Big Data.

As large scale distributed systems such as grids and clouds offer computational services to scientists, consumers and enterprises, there are important issues that must be addressed for large scale distributed systems, such as: performance, resource allocation, efficient scheduling, energy conservation, reliability, cost, availability, quality. Furthermore, due to the cost of electrical power consumption and the environmental impact, energy efficiency in large scale systems is a global IT concern. Effective management of distributed resources is crucial to use effectively the power of these systems and achieve high system performance. Resource allocation and scheduling is a difficult task in large scale distributed systems where there are many alternative heterogeneous computers. The scheduling algorithms must seek a way to maintain a good response time along with energy-efficient solutions that are required to minimize the impact of grid and cloud computing on the environment. Furthermore, the simultaneous usage of computational services of different distributed systems such as clusters, grids, and clouds can have additional benefits such as lower cost and high availability.

Chapter "Scheduling Data-Intensive Workloads in Large-Scale Distributed Systems: Trends and Challenges" presents the backgrond study on the scheduling problems in HPC distribited environments. The detailed taxonomy of the modern scheduling methods and models is defined in detailed, which is one of the best sources of such systematic survey of the recent developments in the domain. The authors present a variety of concepts on HPC systems such as grids and clouds, based on existing or simulated grid and cloud systems, that provide insight into problems solving and we will provide future directions in the grid and cloud computing area. Advanced modelling and simulation techniques are a basic aspect of performance evaluation that is needed before the costly prototyping actions required for complex distributed systems. Queuing network models of large scale distributed systems will be described and analysed and performance metrics will be presented. Complex workloads will be examined including real time jobs and scientific workflows. The second author, who was TS trainer—Prof. Helen Karatza—is a world class expert in the modelling and simulation and scheduling in grids and clouds. She is a Professor in the Department of Informatics at the Aristotle University of Thessaloniki, Greece. Currently, she is a senior member of SCS, IEEE and ACM, and she served as an elected member of the Board of Directors at Large of the Society for Modeling and Simulation International (2009–2011).

Chapter "Design Patterns and Algorithmic Skeletons: A Brief Concordance" is an extension of the training course, which was an introduction into the principles and methods for High-Performance Computing. It made trainees familiar with the tools to develop HPC applications, and form the set of skills for them to understand the pitfalls and subtle details behind optimizing such applications when running them on large distributed infrastructures. In this chapter, the authors addressed one of the challenging problem related to the general topic of the course: how to establish a correspondence between the well-known, accepted design pattern approach and the programmer-oriented functional algorithmic skeleton paradigm. The TS tariner—Dr. Horacio González–Vélez—is vice Chair of the cHiPSet Action and head of the NCI Cloud Computing Center in Dublin, Ireland. He has been recognized with the European Commission ICT award for his efforts on scientific dissemination and the UK NESTA Crucible Fellowship for his inter-disciplinary research on computational science.

The next chapter is on the "Evaluation of Cloud Systems", the corresponding TS course has been provided by Prof. Florin Pop, from University Politehnica of Bucharest, Romania. The chapter covers the fundamental problems related to the evaluation of the clouds usually met in practice. The authors explain how to develop a correct methodology for the evaluation using simulation of Cloud services and components. The CloudSim tool was used in the evaluation section. Florin Pop is an expert in scheduling and resource management (decentralized techniques, re-scheduling), multi-criteria optimization methods, Grid middleware tools and applications development (satellite image processing an environmental data analysis), prediction methods, self-organizing systems, contextualized services in distributed systems.

Science gateways also called portals, virtual research environments or virtual labs form a solution, which offer a graphical user interface tailored to a specific research domain with a single point of entry for job and data management hiding the underlying infrastructure. In the last 10 years quite a few web development frameworks, containerizations, science gateway frameworks and APIs with different foci and strengths have evolved to support the developers of science gateways in implementing an intuitive solution for a target research domain. The selection of a suitable technology for a specific use case is essential and helps reducing the effort in implementing the science gateway by re-using existing software or frameworks.

Thus, a solution for a user community can be provided more efficiently. Additionally, novel developments in web-based technologies and agile web frameworks allow for supporting developers in efficiently creating web-based science gateways.

The topic science gateways and related technologies have gained also importance in the last 10 years for the HPC community. The first time in the history of such solutions, providers of HPC, grid and cloud infrastructures have reported in 2014 that more of their resources have been accessed via science gateways than via command line. The US National Science Foundation (NSF) has recommended a Science Gateway Community Institute for funding, which will provide services starting in July 2016. Additionally, IEEE launched a technical area on science gateways as part of the Technical Committee on Scalable Computing.

Dr. Sandra Gesing from University of Notre Dame (USA) in Chapter "Science Gateways in HPC: Usability Meets Efficiency and Effectiveness" focus on the development of applications over Science gateways. Modeling and simulations, which necessitate HPC infrastructures, are often based on complex scientific theories and involve interdisciplinary research teams. IT specialists support with the efficient access to HPC infrastructures. They design, implement and configure the simulations and models reflecting the sophisticated theoretical models and approaches developed and applied by domain researchers. Roles in such interdisciplinary teams may overlap dependent on the knowledge and experience with computational resources and/or the research domain. Domain researchers are mainly not IT specialists and the requirement to employ HPC infrastructures via command line often forms a huge hurdle for them. Thus, there is the need to increase the usability of simulations and models on HPC infrastructures for the uptake by the user community.

Post-TS Research Results

One of the dreams of the organizers of the Training School in Bucharest was to give to the both trainees and trainers the inspiration to the new research, ideas and developments. Good dreams may come true very quickly. Threfeore, we are very happy to present in the second part of the book chapters with interesting new ideas and results achieved by the school trainees in their research groups in the period of 10 months after the event.

In Chapter "MobEmu: A Framework to Support Decentralized Ad-Hoc Networking", the authors present the simulation toolkit for the opportunistic networks (ON). They mainly focus on the parametrisation of routing process and dissemination of the information and data among the mobile users and resources. The challenge in simulating mobile networks arises from two difficult problems: formalizing mobility features and extracting mobility models. Based on two main mobility models in ONs, namely, models where traces are the results of experiments recording the mobility features of users (location, connectivity, etc.), and synthetic pure mathematical models which attempt to express the movement of devices, the developed simulator allows to conduct experiments on realistic data.

The new challenge in today's data intensive computing is the efficient and secure management of larve volumes of data in the shortest possible time. Recently, some inspirations from the financial virtual markets such as Forex and virtual stock exchanges are discussed in the data mining community. It would allow to keep the physical data at one storage server, while data virtual models can be used for the processing and analysis. Chapter "Virtualisation Model for Processing of the Sensitive Mobile Data" shows the first results of such virtualization. The authors focused on the transmission of the personal fragile data over the obile cloud. The achieved results seems to be very promissing and confirm the promising direction of the further research in this domain.

The computational load of cryptographic procedures in Cloud Computing (CC) systems are crucial to such systems effectiveness. Additionally, assuring Quality of Service (QoS) requirements is possible when the security layer applied to tasks does not interrupt the computing process. Such solutions have to protect both the user data as well as the whole system. They have to support the scalability, multi-tenancy and complexity that CC systems. In Chapter "Analysis of Selected Cryptographic Services for Processing Batch Tasks in Cloud Computing Systems", the authors present a cryptographic service based on blind RSA algorithm and Shamir secret sharing. The service was designed for batch tasks processing by many Virtual Machines (VMs) as working nodes. Blind RSA cryptographic system is used to encrypt the data without actually knowing any details about the tasks content. Shamir secret sharing procedure is proposed in order to assure whether all VMs in the system gave back their shares after deploying the batch of tasks on them or not. Authors provided detailed analysis of proposed cryptographic service. An extensive scalability analysis is presented. Experimental results performed in order to evaluate the proposed model are done on CloudSim simulator.

Acknowledgements

This book project has been inspired and based upon our work from COST Action IC1406 High-Performance Modelling and Simulation for Big Data Applications (cHiPSet), supported by COST (European Cooperation in Science and Technology). We are grateful to all the contributors of this book, especially trainers and trainees as well as the organizers of the cHiPSet Training School 2016 organized in Bucharest in September 2016. We thank also to the rest of the the authors for their interesting proposals of book chapters, their time, efforts and their research results. We also would like to express our sincere thanks to Dr. Ralph Stuebner—cHiPSet Cost Scientific Officer—and all cHiPSet members who have helped us ensure the quality of this volume. We gratefully acknowledge their time and valuable remarks and comments.

Our special thanks go to Prof. Janusz Kacprzyk, editor-in-chief of the Springer's Studies in Big Data Series, Dr. Thomas Ditzinger and all editorial team of Springer Verlag for their patience, valuable editorial assistance and excellent cooperation in this book project.

Finally, we would like to send our warmest gratitude message to our friends and families for their patience, love, and support in the preparation of this volume.

Cracow, Poland
Bucharest, Romania
September 2017

Joanna Kołodziej, cHiPSet Chair
Florin Pop
Ciprian Dobre

Contents

Contributors

Adriana E. Chis Cloud Competency Centre, National College of Ireland, Dublin 1, Ireland

Radu-Ioan Ciobanu University Politehnica of Bucharest, Bucharest, Romania

Ciprian Dobre University Politehnica of Bucharest, Bucharest, Romania

Marc Frincu West University of Timisoara, Timisoara, Romania

Sandra Gesing University of Notre Dame, Notre Dame, USA

Horacio González–Vélez Cloud Competency Centre, National College of Ireland, Dublin 1, Ireland

George-Valentin Iordache Computer Science Department, Faculty of Automatic Control and Computers, University Politehnica of Bucharest, Bucharest, Romania

Bogdan Irimie West University of Timisoara, Timisoara, Romania

Agnieszka Jakóbik Tadeusz Kościuszko Cracow University of Technology, Cracow, Poland

Helen D. Karatza Department of Informatics, Aristotle University of Thessaloniki, Thessaloniki, Greece

Joanna Kołodziej Cracow University of Technology, Cracow, Poland

Radu-Corneliu Marin University Politehnica of Bucharest, Bucharest, Romania

Florin Pop Computer Science Department, Faculty of Automatic Control and Computers, University Politehnica of Bucharest, Bucharest, Romania; National Institute for Research and Development in Informatics (ICI), Bucharest, Romania

Teodora Selea Austria Research Institute, Timisoara, Romania

Adrian Spataru Austria Research Institute, Timisoara, Romania

Georgios L. Stavrinides Department of Informatics, Aristotle University of Thessaloniki, Thessaloniki, Greece

Jacek Tchórzewski Tadeusz Kościuszko Cracow University of Technology, Cracow, Poland; AGH University of Science and Technology Krakow, Cracow, Poland

Alexandru Tudorica Computer Science Department, Faculty of Automatic Control and Computers, University Politehnica of Bucharest, Bucharest, Romania

Mihaela-Andreea Vasile Computer Science Department, Faculty of Automatic Control and Computers, University Politehnica of Bucharest, Bucharest, Romania

Anca Vulpe West University of Timisoara, Timisoara, Romania

Andrzej Wilczyński Cracow University of Technology, Cracow, Poland; AGH University of Science and Technology, Cracow, Poland

Evaluating Distributed Systems and Applications Through Accurate Models and Simulations

Marc Frincu, Bogdan Irimie, Teodora Selea, Adrian Spataru and Anca Vulpe

Abstract Evaluating the performance of distributed applications can be performed by in situ deployment on real-life platforms. However, this technique requires effort in terms of time allocated to configure both application and platform, execution time of tests, and analysis of results. Alternatively, users can evaluate their applications by running them on simulators on multiple scenarios. This provides a fast and reliable method for testing the application and platform on which it is executed. However, the accuracy of the results depend on the cross-layer models used by the simulators. In this chapter we investigate some of the existing models for representing both applications and the underlying distributed platform and infrastructure. We focus our presentation on the popular SimGrid simulator. We emphasize some best practices and conclude with few control questions and problems.

1 Introduction

A **distributed system** (DS) is a collection of *entities* (e.g., process or device) which *communicate* through a communication medium (e.g., wired or wireless network) appearing to end users as a single coherent system. The entities are characterized by autonomicity, programmability, asynchronicity, and failure-proneness, while the communication medium is usually unreliable.

M. Frincu (✉) · B. Irimie · A. Vulpe
West University of Timisoara, Timisoara, Romania
e-mail: marc.frincu@e-uvt.ro

B. Irimie
e-mail: bogdan.irimie90@e-uvt.ro

A. Vulpe
e-mail: anca.vulpe94@e-uvt.ro

T. Selea · A. Spataru
Austria Research Institute, Timisoara, Romania
e-mail: teodora.selea93@e-uvt.ro

A. Spataru
e-mail: florin.spataru92@e-uvt.ro

J. Kołodziej et al. (eds.), *Modeling and Simulation in HPC and Cloud Systems*,
Studies in Big Data 36, https://doi.org/10.1007/978-3-319-73767-6_1

Distributed systems enable a form of parallel computing, namely **distributed computing**, where computation and data is geographically spread, but the applications have parallel tasks processing the same or different parts of the data at the same time.

Designing and testing platforms and application on top of them require therefore specific configurations which due to the nature of the DS may be out of reach of researchers and software engineers. In addition, real-life systems may not offer the complexity and specific setup required by some applications. In situ experiments need therefore to be replaced with a more economic and flexible alternative in which the DS can still be properly modeled. The complexity of DSs (e.g., clouds) make theoretical models not viable as the large number of parameters required to model heterogeneity, dynamism, exascale, and Big Data leads to systems which are too complex to be modeled through mathematics alone.

Simulations offer the fastest path from idea to its testing. They enable users to get preliminary results from partial implementations and to improve their algorithms, applications, and platforms quickly and under various settings and assumptions. They also allow experiments on thousands of configurations fast at no cost and without having to wait in line for available resources. Finally, they allow users to bypass any technical challenges posed by the platform and DS letting them focus on the application itself. Figure 1 depicts the usual simulation flow from idea, experimental setup and model to scientific results.

Despite their advantages, simulations face several challenges including:

- **Validity**: Results obtained through simulations should match or be close to those obtained in real-life experiments. Approximations in any simulation should be quantifiable such that any result would be mapped to its real-life equivalent. For instance, Virtual Machine (VM) boot and stop times which could be ignored in simulations should not impact the outcome of real-life deployment as predicted by the simulation. Furthermore, the accuracy of underlying is essential in validating the experiments. Extensive tests of SimGrid and comparison with other simulators have outlined strange behaviors in the network modeling of simulators such as OptorSim, GridSim, and CloudSim [2].
- **Scalability**: Any simulation should scale with the experiment size to allow fast exploration of scenarios of several orders of magnitude. For instance, the simulation time of a single scenario should not exceed in any case the validation on real-life DSs.
- **Tools**: Simulation results are usually numerically encoded and contain lots of data unreadable in raw format. Furthermore, if multiple scenarios are tested the visual analysis of the raw data is practically impossible. Hence, automated tools for visual analysis are required. In addition, to avoid the time consuming manual generation of hundreds or thousands of experiment settings, automatic generation based on customized parameters is necessary.
- **Applicability**: Simulations should match the user requirements and objectives. Hence, the underlying models should closely match real-life scenarios while the simulation output should match the desired goals.

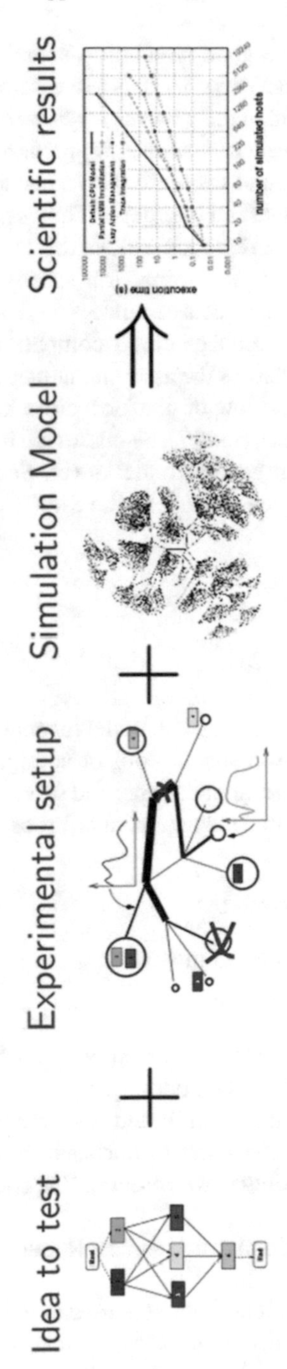

Fig. 1 Simulation flow [1]

Fundamentally, to enable a good simulation sound **models** are required across the simulated platform layers. These models are essential both for the validity and applicability of the simulation. *In this chapter we focus on cloud simulators as clouds are widely researched and new algorithms for numerous problems are developed constantly*. Despite being around for more than a decade clouds have yet to unveil their full potential with Big Data and Internet of Things promising new challenges for clouds. Simulations will play an essential role in driving the next wave of algorithms for topics including Big Data processing, job scheduling in hybrid systems and architectures, and Quality of Service assurance.

The rest of the chapter introduces cloud computing and cloud simulators (cf. Sect. 2), then it moves on to discuss the assessment methods of a distributed application (cf. Sect. 3), gives an overview of platform cross layer models (cf. Sect. 4) and details each of them (cf. Sects. 5, 6 and 7), discusses the importance of simulation data (cf. Sect. 8), concluding with general remarks (cf. Sect. 9). The model layers are presented by mirroring the ones existing in the SimGrid simulator [3] and SchIaaS extension [4].

2 Cloud Computing

According to NIST, "cloud computing is a model for enabling ubiquitous, convenient, on-demand network access to a shared pool of configurable computing resources (e.g., networks, servers, storage, applications, and services) that can be rapidly provisioned and released with minimal management effort or service provider interaction" [5].

Cloud computing is characterized by:

- **On demand access** to storage, computational and network resources, platforms, and applications
- **Broad network access**
- **Pay-per-use policy** varying between resource types and providers. Examples include per hour, per minute, per Gb, per request
- **Resource pooling** with virtually unlimited resources for users
- **Rapid elasticity** which allows users to horizontally (e.g., adding new/removing VMs) or vertically (e.g., adding new/removing VM cores and memory) scale based on demand
- **New programming paradigms** such as MapReduce, Hadoop, NoSQL (Cassandra, MongoDB)
- **Big Data** where the large cloud data centers can now store PBs of data coming from the research community (e.g., astrophysical data, meteorological data) or industry (e.g., social network data, banking data).

All these features are stored in a layered cloud stack (cf. Fig. 2). The Infrastructure as a Service (IaaS) offers direct access to virtualized resources being targeted as

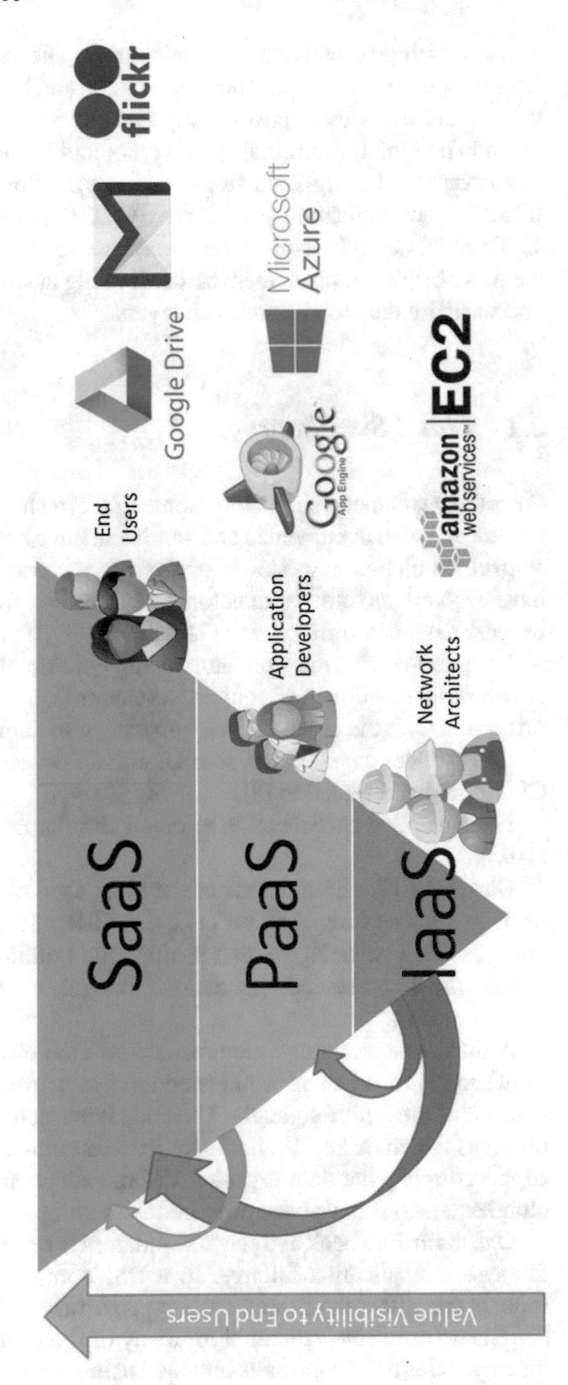

Fig. 2 Cloud stack

network architects and cloud administrators. The Platform as a Service (PaaS) offers OS level functions to application developers, and finally, Software as a Service (SaaS) targets end users by exposing fully fledged applications. The *as a Service* model extends beyond these initial three layers and comprises Data as a Service (DaaS) to emphasize the Big Data stored and accessible in clouds, Network as a Service (NaaS), Communication as a Service (CaaS), and Monitoring as a Service (MaaS).

To simulate such complex environments various simulators have been proposed. Next, we briefly compare them before looking at what a good simulation should offer and what are the cloud simulation layers.

2.1 Cloud Simulators

Over the years many simulators, some of them short lived, others widely used in literature and well documented and validated, have been devised. Many of them started as grid simulators have slowly evolved in generic cloud simulators [3]. Others [6] have evolved into cloud simulators by borrowing from other the underlying models, or were designed from scratch [7].

Despite many of the simulators being generic, there are also corner cases where custom built solutions are required. As example, IoT applications are becoming better integrated with the cloud and they give birth to another layer between the cloud and the IoT devices called *fog*. A simulator that addresses this corner case and based on CloudSim is presented in [8].

For a detailed classification of cloud simulators we direct the readers to [9] and [10].

Currently, CloudSim is one of the most used cloud simulator with many related projects such as CloudAnalyst [11], CloudReports [12, 13] expanding its functionality. Its main advantage is that it offers the building blocks for modeling complex cloud infrastructure and applications that run on top of them. The entire code is written in Java.

A different simulator which focuses on DSs but which has support for clouds is SimGrid [3]. It relies on well tested models across the simulated system stack and extensions are built frequently. The code is written in C but has Java, Ruby, and Lua bindings which makes it suitable for a wider audience. SimGrid does not offer cloud support directly but does expose a VM migration and execution model. Based on it, cloud extensions have been proposed with an updated list available at [14].

One main drawback of today's simulators is that they do not have accurate models for system/application failures. In a DS, components fail all the time and those failures can affect the overall system performance. Another major drawback is that there is no extensive validation for many of the simulators and thus there can be big discrepancies between results obtained using the simulators and experiments on real cloud infrastructures due to poor or simplistic models.

3 Assessment of Applications

When assessing cloud applications users are usually interested in its **correctness** and **performance**. Each is quantified through various metrics depending on the application objectives. Hence correctness can be modeled through the absence of crashes, race conditions or deadlocks. Performance indicators usually address makespan, throughput, energy consumption or running costs. A comprehensive overview of client side objectives is given in our previous work [15].

Due to the complexity of the environment the **correction study** for a cloud application relies on model-checking as it allows an exhaustive automated exploration of the state space to identify issues. Instead, **performance studies** rely on simulations to test prototype applications on system models in scenarios unavailable on real-life systems and where math is insufficient to understand the behavior of the applications. An alternative to simulations could be emulations which rely on testing real-life applications on synthetic systems.

In simulations a key requirement is the **reproducibility** of simulation results which allows users to rerun the same experimental setup described in a paper to benchmark on a different data set or to compare with a different approach. However, one of the main problems in literature is the lack of sufficient details and that of publicly available source codes on which the experiments were based on.

Another important aspect when running simulations is to have access to **standard tools** which users can learn quickly without having to code their own software or to learn several simulators for different simulation objectives. In practice, there are lots of custom made shot lived simulators which do not provide insight on the models used to simulate the cloud systems or make assumptions which may not be valid. Furthermore, their validity has not been thoroughly tested despite being used in many research papers [1].

By having failures inserted in the simulations, we can observe how application behaves under abnormal conditions and establish limits in the amount of failures that our application can cope with.

4 Modeling Layers

When designing a cloud simulation and simulator, the entire cloud stack needs to be considered to ensure a proper representation of the real-life environment. For each of the layers, accurate models need to be implemented and validated on large amounts of data to ensure the validity of the subsequent simulations.

Hence, the following minimal list of models should be implemented:

1. Bare metal models
    - Hardware models for CPU, network, memory;

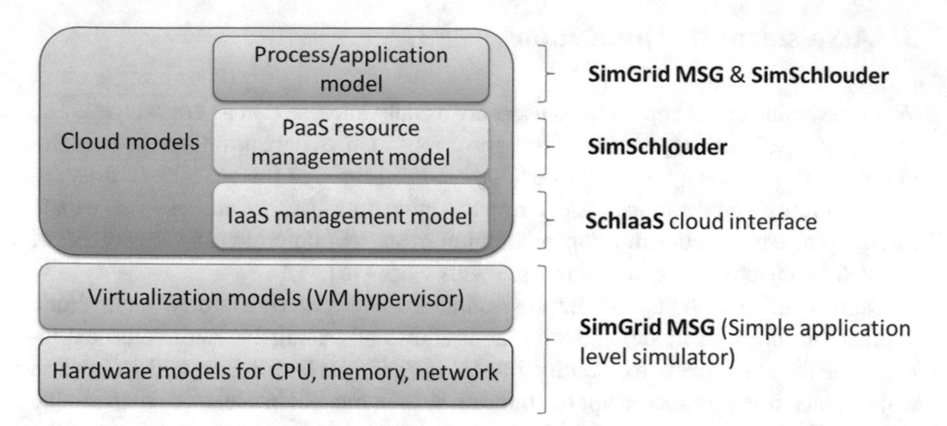

Fig. 3 Simulation stack with the equivalent SimGrid core modules and extensions

2. Virtualization models
   - Models for hypervisors;
3. Cloud models
   - IaaS level management similar to existing providers' APIs;
   - OS level (PaaS) models for proper abstractions and resource management;
   - Models for simplistic yet accurate process/application representation.

Next, we give an overview of each of them by providing implementation examples with reference to the SimGrid simulator. Figure 3 depicts the simulation stack of SimGrid with references to the SimGrid internal modules (i.e., SimGrid MSG) or existing extensions (i.e., SimSchlouder, SchIaas – see Sects. 7.1 and 7.2 for more details).

5 Hardware Model

At the hardware model level simulators need to consider CPU and network.

CPU processing speed s is usually expressed in flops (floating point operations per second) which means that processes running on them should be specified in terms of required floating point operations r. At this level modeling is trivial since in order to get the execution time of a process we simply have to compute r/s.

In resource sharing environments it is possible to model CPU sharing and to introduce CPU availability traces to model the fluctuation of the CPU speed.

However, modern architecture are usually parallel machines with multiple cores and processors. These architectures are usually modeled by having an array a which describes the number of floating point operations that each processor has to execute and a matrix B which describes the communication pattern [16]. This enables the modeling of:

- Fully parallel tasks: $a \neq 0$ and $B = 0$;
- Data redistribution tasks: $a = 0$ and $B \neq 0$;
- Tasks with communication: $a \neq 0$ and $B \neq 0$.

The model can be further extended to account for inter-processor cache sharing, memory, and compiler/OS particularities.

Network modeling is more complex and needs to account for latency, bandwidth sharing, and TCP congestion in order to obtain realistic simulations. Several simulation models exist in literature:

- **Delay-based** models: are the simplest network models. They allow the modeling of communication time through statistical models, constants (e.g., latency), and geographical coordinate systems to account for geographic proximity. The motivation behind these models is that end-to-end delay greatly affects the performance of applications running on the network [17]. One of their main drawbacks is that they ignore network congestion and assume large bisection bandwidth (i.e., the available bandwidth between endpoints).
- **Packet level** models: capture the behavior and interaction of individual packets through complex models. Examples of simulators taking this approach include GTNetS [18], NS2 [19] which simulate the entire protocol stack.
- **Flow level** models: simulate the entire communication flow as a single entity $T_{i,j}(S) = L_{i,j} + S/B_{i,j}$, where S represents the message size, $L_{i,j}$ is the latency between endpoints i and j, and $B_{i,j}$ represents the bandwidth. The model assumes a steady-state and bandwidth sharing each time a new flow appears or disappears. Given a flow ϕ_k and the capacity of the link C_j the constraint is to have $\sum_k \phi_k < C_j$. Several algorithms including Max-Min fairness, Vegas, and Reno exist. In case of Max-Min fairness the objective is to $\max \min(\phi_k)$ with the equilibrium reached when increasing any flow ϕ_l decreases a given flow ϕ_k. As such it tries to give a fair share to all flows sharing the link.

Besides models for simulating data flows the hardware model also comprises of models of the topology. In SimGrid for instance, a DS is represented as a static

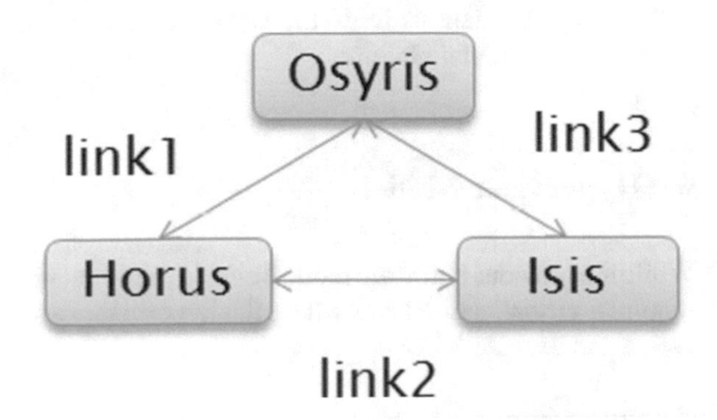

Fig. 4 Simple DS topology

topology as seen in the following simple example[1] which defines an autonomous system with full routing comprised of three machines linked together as in Fig. 4. The hardware model in Simgrid is part of the core MSG module and can be used when modeling any DS.

```
<platform version="3">
 <AS  id="AS0"  routing="Full">
  <host id="Horus" core="4" power="8095000000"
   availability_file="horus_avail.trace"
   state_file="horus.state" />
  <host id="Osyris" core="4" power="8095000000"/>
  <host id="Isis" core="4" power="8095000000"/>

  <link id="link1" bandwidth="125000000"
   latency="0.000100" bandwidth_file="link1.bw"
   latency_file="link1.lat"/>
  <link id="link2" bandwidth="125000000"
   latency="0.000100"/>
  <link id="link3" bandwidth="125000000"
   latency="0.000500"/>

  <route src="Horus" dst="Osyris"><link_ctn id="link1"/>
  </route>
  <route src="Horus" dst="Isis"><link_ctn id="link2"/>
  </route>
  <route src="Osyris" dst="Isis"><link_ctn id="link3"/>
  </route>
 </AS>
</platform>
```

where power is in flops, latency is in seconds, and bandwidth is in bytes/second. While the topology is fixed there is the option to define traces for CPU, bandwidth and latency fluctuations (e.g., the cases of horus host and link2), and availability periods (e.g., simulate failures).

6 Hypervisor Model

While the previous hardware models enable the simulation of DS such as grids, clouds require a virtualization layer where the hypervisor (e.g., Xen, KVM, VMWare) can create and execute VMs.

[1]Full documentation available at: http://simgrid.gforge.inria.fr/simgrid/3.12/doc/platform.html.

To enable a seamless transition from simulation to real-life deployment models should mimic the real systems. Hence the user level API should match that of existing hypervisors with functions for starting, stopping, pausing, and resuming VMs.

Simulators such as SimGrid implement [20] such APIs and offer models for live migration as well.

To enable **VM management** two resource constraint problem need to be solved, at physical level, and at virtualized level. In SimGrid for instance, VMs are seen as an ordinary task executed on the physical machine. Basically, to place VMs along side regular tasks the simulator first computes the share of the host for each of them. Then, for each VM it computes the shares of tasks running on them using the allocated shared by the host as maximum. For instance if a host has a capacity C and there are 2 VMs and one task allocated to it it first solves the constraint $S_{VM_1} + S_{VM_2} + S_t < C$, where S_* represents the share of the host to be allocated. Second, once S_{VM_1} and S_{VM_2} are determined, assuming VM_1 will execute 2 tasks and VM_2 one task, it solves the constraints $S_{t_1} + S_{t_2} < S_{VM_1}$ and $S_{t_3} < S_{VM_2}$. In addition, task priorities and VM CPU usage capping can be specified.

Once a hypervisor model is in place, the **live migration** of VMs needs to be modeled too. This capability is at the core of activities involving system maintenance, load balancing, energy efficiency, reconfiguration and fault tolerance. Some simulators such as CloudSim let users specify the migration time but this approach is simplistic. Default live migration policies can be overridden hence allowing for some flexibility and testing of new algorithms.

In SimGrid, the precopy live migration **algorithm** is implemented as part of the core MSG, however in literature other well-known algorithms such as post copy and hybrid exist. A detailed overview and analysis of their performance is given in [21]. The reason for implementing the precopy algorithm is its popularity among well-known hypervisors such as Xen, KVM, and VMWare.

- Precopy: the algorithm iteratively copies memory pages of the VM from the source host to the destination. First, it copies all memory pages. At subsequent steps it copies only the modified pages, and repeats this step until the number of modified pages is small enough. At this stage it stops the VM and copies the remaining dirty pages to the destination. Finally, it restarts the VM at the destination. The entire process takes from few ms to seconds. The algorithm is reliable and robust as the entire process can be rolled back if the migration fails.
- Postcopy: the algorithm first stops the VM and then copies using demand and prepaging techniques over the network. First, the VM makes some initial preparation of resources. Then, the VM is stopped and the execution states are transferred and switched on at the destination host to resume the VM. During this phase the VM is down. After the states have been transferred and the VM has resumed the memory page will be copied. In this algorithm the transferred VM will start immediately but will suffer from performance penalties from network page faults. The performance of this algorithm is highly dependent on the workload and hence choosing it requires a deep analysis with different workloads.

- Hybrid: the algorithm is a special version of postcopy where a limited number of precopy stages are applied a priori. The algorithm is useful in cases where we want to balance the reliability of precopy with speed of postcopy.

Depending on whether or not users want to investigate live migration algorithms simulators can offer extensible constructs to enable their validation by relying on the hardware models.

7 Cloud Model

With a virtualization model in place simulators can be augmented with support for cloud models. These models should mimic the layered cloud architecture at IaaS and PaaS with support for running applications at each one. Simulators should be generic and extensible to allow the insertion of new cloud engines. Popular cloud IaaS models include the Amazon EC2 model of instances and billing. The complexity and level of Amazon EC2 services has enabled a vast collection of EC2 compatible APIs in various cloud software platforms such as Eucalyptus, OpenStack, and OpenNebula to name a few.

Contrary to the hypervisor layer, in the cloud layer users handle instances not VMs. These instances have several characteristics including type and billing model, and are automatically placed on hosts by the hypervisor. In SimGrid, users can access cloud IaaS APIs through the SchIaas extension, while PaaS level resource management for bag-of-tasks and workflow applications can be handled through SimSchlouder.

7.1 Infrastructure Model

In SimGrid, the cloud topology including compute and storage services, instance types, instance images, and the physical infrastructure to host the VMs algorithms is defined in a file similar to the simple example below:

```
<clouds version="1">
 <cloud  id="myCloud">
  <storage id="myStorage"
    engine="org.simgrid.schiaas.engine.storage.rise.Rise">
    <config controller="Horus"/>
  </storage>

  <compute
    engine="org.simgrid.schiaas.engine.compute.rice.Rice">
```

```
<config controller="Horus" image_storage="myStorage"
   image_caching="PRE inter_boot_delay="10"/>

<instance_type id="small" core="1" memory="1000"
 disk="1690"/>
<instance_type id="medium" core="2" memory="1000"
 disk="1690"/>
<instance_type id="large" core="4" memory="1000"
 disk="1690"/>

<image id="myImage" size="1073741824"/>

<host id="Osyris"/>
<host id="Isis"/>
  </compute>
 </cloud>
</clouds>
```

The IaaS model usually has two views. The cloud client view available to end users where compute instances and storage can be handled; and the cloud provider view where cloud IaaS administrators handle VM to host placement and other cloud infrastructure management activities. In SimGrid, the provider view is handled by default by the RICE (Reduced Implementation of Compute Engine) and RISE (Reduced Implementation of Storage Engine) engines part of the SchIaaS extension (cf. Fig. 5).

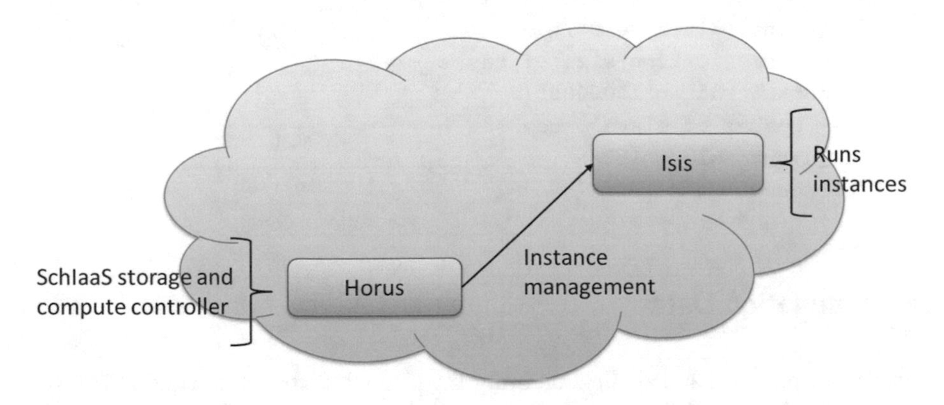

Fig. 5 Simple IaaS model in SchIaaS

7.2 *Platform Model*

At PaaS level simulators usually provide functionality and models for simulating application execution. Resource management for simulating bag-of-tasks and workflow applications is an example of such functionality.

At this level users can test scheduling algorithms on various applications, and cloud and infrastructure topologies by relying on the simulator models for computation, communication, virtualization, and cloud. For simulators such as SimGrid this is the where users take advantage of the full simulator stack to propose new models for cloud resource management. SimSchlouder is one such extension which offers basic task to VM scheduling heuristics.

7.3 *Application Model*

To simulate applications we require simplistic yet comprehensive models for them. Required information should be mapped on the underlying models, namely on the computation and communication models. The following simple example specifies a process that will spawn a job with 10 tasks with predefined size in floating point operations and communication size in bytes.

```
<process host="Horus" function="cloud.schiaas.Master">
  <!-- Number of tasks -->
  <argument value="10"/>
  <!-- Computation size of tasks -->
  <argument value="5e10"/>
  <!-- Communication size of tasks -->
  <argument value="1000000"/>
  <!-- Number of slave processes -->
  <argument value="10"/>
</process>
```

8 Simulation Data

Traces for platform and application bring the simulated application and DS closer to the behavior of real-life systems. **Traces** can be either **synthetic** or from **real-life** systems. Synthetic traces are based on statistical analysis of real-life systems and capture variations which on real traces may not be visible. A detailed overview of synthetic data and how to generate it for DS is given in [22]. A comparative study – from more than 2 decades ago – between the two trace types has outlined no significant differences in algorithm behavior [23].

There are many large trace sources from companies like Wikipedia [24], Google [25] as well as traces from various parallel [26] and grid systems [27]. Despite their advantages, one downside is that a large portion of the simulation time is spent reading the traces from disk as those traces can have hundreds of GB in size. Furthermore, different trace sources can use different formats for storing data hence content parsers need to be customized based on each source.

Nevertheless, traces are a suitable choice when the goal is to build a simulation as close as possible to a real-life scenario.

9 Conclusion

In this chapter we have emphasized the importance of simulators and simulation models. Real-life systems require a huge amount of effort to configure the environment (application and platform), to run tests, and to analyze results.

In contrast, simulators allow users a fast and reliable method to evaluate applications ran on a specific platform. Over the years, the diversity of simulators has increased leading to general purpose and specialized simulators on energy efficiency, network modeling, network-aware scheduling, workload planning, resource allocation, service brokering, storage modeling (cf. Table 1).

Building cloud simulations and simulators consists in implementing bare metal models, virtualization models, and cloud models.

The bare metal model is represented by hardware model level. Here has to be taken in consideration CPU and network. Modeling CPU is a simpler than network modeling which has to take into account latency, bandwidth, and TCP congestion.

At virtualization level, the hypervisor model can create and execute VMs. The most well known algorithms implemented at this level are precopy, postcopy and hybrid.

Cloud models should mimic the layered cloud architecture at the IaaS and PaaS layers. Simulators need to be generic and extensible to allow extensions and customized behavior. The IaaS model is composed of two main views: client view and provider view. The PaaS level simulators provide functionality and models for simulating application execution. The application model is mapped on the underlying models to enable users to take full advantage of the simulation environment.

Simulation data is represented either by real-life data or by synthetic data. Synthetic data is based on statistical analysis of real-life systems. It can captures variations which may not be visible on real traces.

In conclusion, with the increase complexity of DSs we expect simulators to play a crucial role in both research and development industry by enabling applications to be tested in scenarios not covered by the limitations of real-life systems.

Table 1 Recommended cloud simulators depending on user priorities [10]

Scenario	CloudSim	Green Cloud	iCanCloud	Network CloudSim	Cloud Analyst	GoundSim	CDOSim	MDCSim	GDCSim	SPECI	BigHouse	TeachCloud	SimGrid
Simple state sharing problem/high number of Nodes (>10k Nodes)	✓	×	×	×	×	✓	×	×	×	✓	✓	✓	✓
Energy efficiency/energy aware scheduling	×	✓	✓	×	×	×	×	×	×	×	×	×	×
Energy efficiency/cooling	×	×	×	×	×	×	×	×	✓	×	×	×	×
High availability/fault tolerance	✓	✓	×	×	×	✓	×	×	×	×	×	×	✓
Network modeling and network-aware scheduling	×	✓	×	✓	×	×	×	✓	×	✓	×	×	✓
Workload planning/evaluation	✓	✓	×	×	✓	×	×	✓	×	×	✓	×	✓
Workflow Modeling	✓	✓	×	×	×	✓	×	×	×	×	×	×	✓
Resource allocation	✓	✓	✓	✓	✓	×	×	✓	×	×	✓	✓	✓
Service brokering	✓	×	✓	×	✓	×	✓	×	×	×	×	×	×
Storage modeling	×	×	✓	×	×	×	×	×	×	×	✓	✓	✓
GUI/easy Use	×	×	✓	×	✓	×	×	×	×	✓	×	✓	✓
High request/job load (>10k requests)	✓	✓	✓	✓	✓	✓	✓	✓	×	✓	✓	✓	✓
MapReduce application modeling/data replication	✓	✓	×	×	×	×	×	×	×	×	✓	✓	✓

Acknowledgements This chapter is based upon work from COST Action IC1406 High-Performance Modelling and Simulation for Big Data Applications (cHiPSet), supported by COST (European Cooperation in Science and Technology).

Additionally, the first author has been invited as a trainer to the cHiPSet training school "New trends in modeling and simulation in HPC system" held in Bucharest in September 21–23, 2016 and has been supported by the IC1406 Horizon 2020 grant. His work has also been partially funded by a grant of the Romanian National Authority for Scientific Research and Innovation, CNCS/CCCDI - UEFISCDI, project number PN-III-P3-3.6-H2020-2016-0005, within PNCDI III. The work of the second author has been partially funded by the EU H2020 VI-SEEM project under contract no. 675121. The work of the third and forth authors has been partially funded by the EU H2020 CloudLightning project under grant no. 643946.

References

1. Martin, Q., et al.: Simgrid 101: Getting started to the simgrid project (Jan 2015). http://simgrid.gforge.inria.fr/tutorials/simgrid-101.pdf
2. Simgrid Models. Getting started with simgrid models (2016). http://simgrid.gforge.inria.fr/tutorials/surf-101.pdf
3. Casanova, H., Giersch, A., Legrand, A., Quinson, M., Suter, F.: Versatile, scalable, and accurate simulation of distributed applications and platforms. J. Parallel Distrib. Comput. **74**(10), 2899–2917 (2014)
4. Julien, G., et al.: Iaas simulation upon simgrid (2015). http://schiaas.gforge.inria.fr/
5. NIST. Cloud computing (2016). https://www.nist.gov/itl/cloud-computing
6. Calheiros, R.N., Ranjan, R., Beloglazov, A., De Rose, C.A.F., Buyya, R.: Cloudsim: a toolkit for modeling and simulation of cloud computing environments and evaluation of resource provisioning algorithms. Softw. Pract. Exper. **41**(1), 23–50 (2011)
7. Núñez, A., Vázquez-Poletti, J.L., Caminero, A.C., Castañé, G.G., Carretero, J., Llorente, I.M.: icancloud: a flexible and scalable cloud infrastructure simulator. J. Grid Comput. **10**(1), 185–209 (2012)
8. Gupta, H., Vahid Dastjerdi, A., Ghosh, S.K., Buyya, R.: ifogsim: a toolkit for modeling and simulation of resource management techniques in internet of things, edge and fog computing environments. *CoRR*, abs/1606.02007 (2016)
9. Ahmed, A., Sabyasachi, A.S.: Cloud computing simulators: a detailed survey and future direction (Feb 2014)
10. Sharkh, M.A., Kanso, A., Shami, A., Öhlén, P.: Building a cloud on earth: a study of cloud computing data center simulators. Comput. Netw. **108**, 78–96 (2016)
11. Wickremasinghe, B., Calheiros, R.N., Buyya, R.: Cloudanalyst: a cloudsim-based visual modeller for analysing cloud computing environments and applications. In: 2010 24th IEEE International Conference on Advanced Information Networking and Applications, pp. 446–452 (Apr 2010)
12. Samimi, P., Teimouri, Y., Mukhtar, M.: A combinatorial double auction resource allocation model in cloud computing. Inf. Sci. **357**, 201–216 (2016)
13. Sá, T.T., Calheiros, R.N., Gomes, D.G.: CloudReports: An Extensible Simulation Tool for Energy-Aware Cloud Computing Environments, pp. 127–142. Springer International Publishing, Cham (2014)
14. Simgrid Cloud. Virtualization/cloud abstractions in simgrid (2016). http://simgrid.gforge.inria.fr/contrib/clouds-sg-doc.php
15. Frîncu, M.E., Genaud, S., Gossa, J.: Client-side resource management on the cloud: survey and future directions. IJCC **4**(3), 234–257 (2015)
16. Hunold, S., Casanova, H., Suter, F.: From simulation to experiment: a case study on multiprocessor task scheduling. In: 2011 IEEE International Symposium on Parallel and Distributed Processing Workshops and Phd Forum (IPDPSW), pp. 665–672 (2011)

17. Ghorbanzadeh, M., Abdelhadi, A., Clancy, C.: Delay-Based Backhaul Modeling, pp. 179–240 (2017)
18. Riley, G.F.: Large-scale network simulations with gtnets. In: Simulation Conference, 2003. Proceedings of the 2003 Winter, vol. 1, pp. 676–684 (2003)
19. ISI. The network simulator (Nov 2016). http://www.isi.edu/nsnam/ns/
20. Hirofuchi, T., Lebre, A., Pouilloux, L.: Simgrid vm: virtual machine support for a simulation framework of distributed systems. IEEE Trans. Cloud Comput. (99):1–1 (2015)
21. Shah, S.A.R., Jaikar, A.H., Noh, S.Y.: A performance analysis of precopy, postcopy and hybrid live vm migration algorithms in scientific cloud computing environment. In: 2015 International Conference on High Performance Computing Simulation (HPCS), pp. 229–236 (2015)
22. Feitelson, D.G.: Workload Modeling for Computer Systems Performance Evaluation. Cambridge University Press, Cambridge (2015)
23. Lo, V., Mache, J., Windisch, K.: A Comparative Study of Real Workload Traces and Synthetic Workload Models for Parallel Job Scheduling, pp. 25–46. Springer, Berlin (1998)
24. Urdaneta, G., Pierre, G., van Steen, M.: Wikipedia workload analysis for decentralized hosting. Elsevier Comput. Netw. **53**(11), 1830–1845 (2009)
25. Google. Google traces (2016). https://github.com/google/cluster-data
26. Feitelson, D.: Parallel workload archive (2016). http://www.cs.huji.ac.il/labs/parallel/workload/
27. Iosup, A., et al.: Grid workload archive (2016). http://gwa.ewi.tudelft.nl/

Scheduling Data-Intensive Workloads in Large-Scale Distributed Systems: Trends and Challenges

Georgios L. Stavrinides and Helen D. Karatza

Abstract With the explosive growth of big data, workloads tend to get more complex and computationally demanding. Such applications are processed on distributed interconnected resources that are becoming larger in scale and computational capacity. Data-intensive applications may have different degrees of parallelism and must effectively exploit data locality. Furthermore, they may impose several Quality of Service requirements, such as time constraints and resilience against failures, as well as other objectives, like energy efficiency. These features of the workloads, as well as the inherent characteristics of the computing resources required to process them, present major challenges that require the employment of effective scheduling techniques. In this chapter, a classification of data-intensive workloads is proposed and an overview of the most commonly used approaches for their scheduling in large-scale distributed systems is given. We present novel strategies that have been proposed in the literature and shed light on open challenges and future directions.

Keywords Big data · Data-intensive applications · Gang scheduling · Workflow scheduling · Bag-of-Tasks scheduling · Data locality · Time constraints · Fault tolerance · Energy efficiency

1 Introduction

The ever-increasing momentum of the Internet of Things, the rapid pace of technological advances in mobile devices and cloud computing, as well as the explosive growth of social media, produce an overwhelming flow of data of unprecedented volume and variety at a record rate. Such data are commonly referred to as *big data* and are characterized by the following attributes: (a) volume, i.e. they consist of very

G. L. Stavrinides (✉) · H. D. Karatza
Department of Informatics, Aristotle University of Thessaloniki,
54124 Thessaloniki, Greece
e-mail: gstavrin@csd.auth.gr

H. D. Karatza
e-mail: karatza@csd.auth.gr

J. Kołodziej et al. (eds.), *Modeling and Simulation in HPC and Cloud Systems*,
Studies in Big Data 36, https://doi.org/10.1007/978-3-319-73767-6_2

large datasets, (b) variety, i.e. they comprise diverse structured and unstructured data of various types and (c) velocity, i.e. the data are generated and streamed at staggering speeds [16, 31]. Computationally intensive applications are employed in a wide spectrum of domains such as healthcare, science, engineering, business and finance, in order to unleash the power of big data, extract useful knowledge and gain valuable insights [51].

The advent of big data has called for a paradigm shift in the computer architecture, and consequently the applications, required for their effective processing. Data-intensive applications are typically processed on interconnected computing resources that are geographically distributed, encompass various heterogeneous components, utilize virtualization, feature multi-tenancy and are able to scale up in the foreseeable future. Computer clusters, computational grids and clouds are examples of such platforms [13]. Furthermore, novel hybrid approaches have emerged, such as *fog computing*, which extends the cloud computing paradigm by bringing data processing at computational resources at the edge of the network, closer to where the data are generated, while sending selected data to the cloud for historical analysis and long-term storage [4, 9].

Data-intensive applications may have different degrees of parallelism and must effectively exploit data locality. Furthermore, they may also impose several Quality of Service (QoS) requirements, such as time constraints and resilience against failures, as well as other objectives, like energy efficiency. These features of the workloads operating on big data, as well as the characteristics of the computing resources required to process them, present major challenges that require the employment of effective *scheduling algorithms*. Due to their inherent complexity, the performance of such algorithms is usually evaluated by simulation, rather than by analytical methods. Analytical modeling is difficult and often requires several simplifying assumptions that may have an unpredictable impact on the results [45].

This chapter is organized as follows: Sect. 2 gives a definition of the scheduling problem in large-scale distributed systems, as well as some of the most important scheduling objectives. In Sect. 3, a classification of data-intensive workloads is proposed, according to their degree of parallelism. An overview of the most widely used strategies for the scheduling of each class of data-intensive applications in large-scale distributed systems is given. Section 4 presents some of the major challenges of data-intensive workload scheduling, covering topics such as data locality awareness, timeliness, fault tolerance and energy efficiency. Furthermore, novel strategies that have been proposed in the literature are presented in Sect. 5. Finally, Sect. 6 concludes this chapter, shedding light on open challenges and future research directions.

2 Scheduling Problem

In its general form, the scheduling problem in large-scale distributed systems concerns the mapping of a set of application tasks $V = \{n_1, n_2, \ldots, n_N\}$ to a set of processors $P = \{p_1, p_2, \ldots, p_Q\}$, in order to complete all tasks under the specified constraints (e.g. complete each task within its deadline) [5, 20]. In this general form, the scheduling problem has been shown to be NP-complete [14].

2.1 Scheduling Objectives

Some of the parameters that characterize a task $n_i \in V$ are shown in Fig. 1. These parameters are:

- *arrival time* $a(n_i)$: it is the time at which the task arrives at the system.
- *start time* $s(n_i)$: it is the time at which the task starts its execution.
- *finish time* $f(n_i)$: it is the time at which the task finishes its execution.
- *deadline* $d(n_i)$: it is the time before which the task should finish its execution.

Based on the above parameters, some of the most commonly used scheduling objectives in large-scale distributed systems are:

(a) To minimize the *average response time* $\overline{R}$ of the tasks $n_i \in V$, where $\overline{R}$ is given by:

$$\overline{R} = \frac{1}{N} \sum_{n_i \in V} R(n_i) \tag{1}$$

where $R(n_i) = f(n_i) - a(n_i)$ and N is the number of tasks in V.

(b) To minimize the *makespan* (i.e. total execution time) M of the tasks $n_i \in V$, where M is defined as:

$$M = \max_{n_i \in V} \{f(n_i)\} - \min_{n_i \in V} \{s(n_i)\} \tag{2}$$

(c) To maximize the *task guarantee ratio* TGR of the tasks $n_i \in V$, where TGR is given by:

$$TGR = \frac{1}{N} \sum_{n_i \in V} guar(n_i) \tag{3}$$

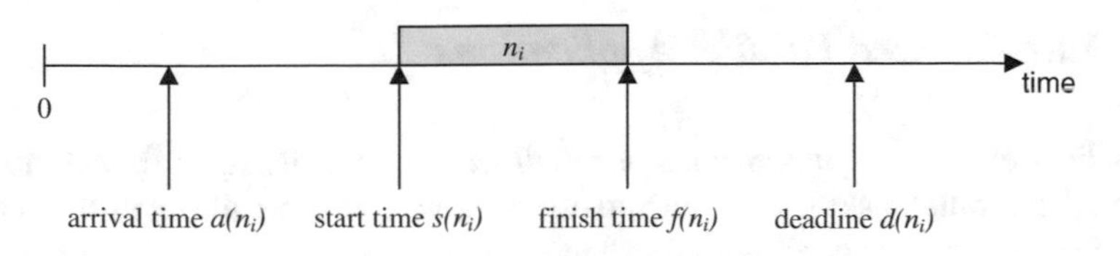

Fig. 1 Typical parameters that characterize a task of an application submitted for execution in a large-scale distributed system

where

$$guar(n_i) = \begin{cases} 1 & \text{if } f(n_i) \leq d(n_i) \\ 0 & \text{otherwise} \end{cases} \quad (4)$$

(d) To minimize the *average tardiness* $\overline{T}$ of the tasks $n_i \in V$, where $\overline{T}$ is defined as:

$$\overline{T} = \frac{1}{N} \sum_{n_i \in V} T(n_i) \quad (5)$$

where

$$T(n_i) = \begin{cases} f(n_i) - d(n_i) & \text{if } f(n_i) > d(n_i) \\ 0 & \text{otherwise} \end{cases} \quad (6)$$

3 Data-Intensive Workloads in Large-Scale Distributed Systems

The data-intensive applications scheduled for execution in large-scale distributed systems, typically consist of numerous component tasks. At the one end of the spectrum, the tasks require frequent communication with each other during their execution. At the other end of the spectrum, the component tasks do not require any communication and are completely independent. Between these two ends, is the case where communication is required between the component tasks of an application, but only before or after their execution. Consequently, data-intensive workloads in large-scale distributed systems can be classified into the following categories:

- *fine-grained parallel applications*,
- *coarse-grained parallel applications* and
- *embarrassingly parallel applications*.

In the following paragraphs, each class of data-intensive applications is presented in more detail and their corresponding, most widely used scheduling heuristics are analyzed.

3.1 Fine-Grained Parallel Applications

An application features *fine-grained parallelism* when it consists of frequently communicating parallel tasks. A proven and effective way to schedule such applications is *gang scheduling*. According to this approach, the parallel tasks of an application form a *gang* and are scheduled and executed simultaneously on different processors. Hence, all of the tasks of the application start execution at the same time. This way,

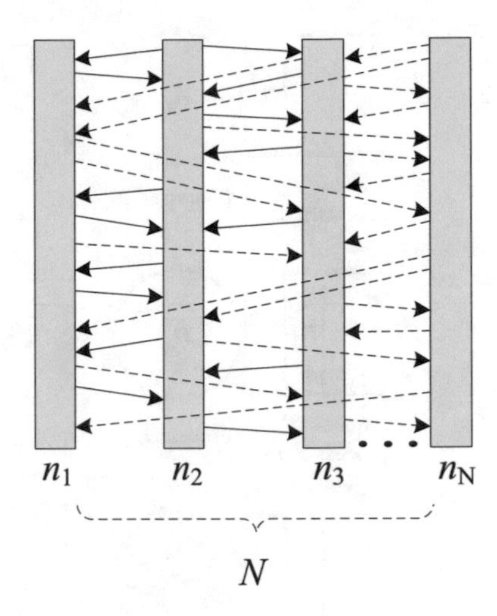

Fig. 2 An example of a fine-grained parallel application. The frequently communicating tasks of the application form a gang of N parallel tasks. The communication between the tasks is depicted with arrows

the risk of a task waiting to communicate with another task that is currently not running is avoided. The task with the largest execution time determines the execution time of the gang. An example of a gang with N parallel tasks is shown in Fig. 2.

Consequently, gang scheduling facilitates the synchronization between the component tasks of a fine-grained parallel application. Without this technique, the synchronization of the component tasks would require more context switches and thus additional overhead. On the other hand, in order to utilize gang scheduling, the number of available processors must be greater than or equal to the number of parallel tasks of the application. Furthermore, due to the requirement that all of the tasks of a gang must start execution at the same time, there may be times at which some of the processors are idle, even with tasks waiting in their respective queues. Specifically, a task at the head of the queue of an idle processor may be waiting for the other tasks of its gang, which may not be able to start execution at the particular time instant [42]. This situation is depicted in Fig. 3.

3.1.1 Gang Scheduling Policies

The two most widely used gang scheduling policies are the *Adapted First Come First Served (AFCFS)* and *Largest Gang First Served (LGFS)* strategies.

Adapted First Come First Served (AFCFS)

This method is an adapted version of the First Come First Served (FCFS) scheduling heuristic, according to which the gang that arrived first, has the highest priority for execution. A gang starts execution when its tasks are at the head of their assigned queues and the respective processors are idle. When there are not enough idle processors for a gang with a large number of parallel tasks waiting at the front of their assigned queues, a smaller gang with tasks waiting behind those of the larger gang can start execution. This technique is also referred to as *backfilling* [18].

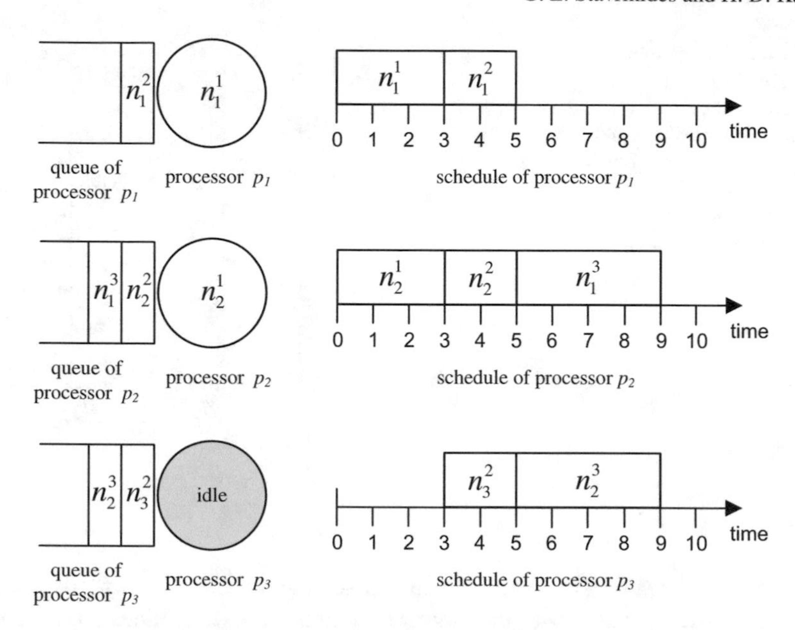

Fig. 3 Example of gang scheduling in a system with three processors p_1, p_2 and p_3. The first gang consists of the tasks n_1^1 and n_2^1, scheduled on processors p_1 and p_2, respectively. The second gang consists of the tasks n_1^2, n_2^2 and n_3^2, scheduled on processors p_1, p_2 and p_3, respectively. The third gang consists of the tasks n_1^3 and n_2^3, scheduled on processors p_2 and p_3, respectively. It can be observed that the processor p_3 remains idle during the execution of the tasks n_1^1 and n_2^1 of the first gang. This is due to the fact that the task n_3^2 at the head of its queue cannot start execution, because according to the gang scheduling technique, it must start execution at the same time as the other tasks of its gang, n_1^2 and n_2^2, which are scheduled on the other processors that are currently busy

The major drawback of this scheduling policy is that it tends to favor smaller gangs, which leads to greater response times for larger gangs. In order to overcome this issue, various techniques have been proposed in the literature, such as the employment of a *bypass count* parameter [25] and the utilization of *task migrations* [30]. The first method, counts for each gang the number of gangs that bypassed it, due to an insufficient number of idle processors. When the bypass count of a gang reaches a specified threshold, it gets the highest priority for execution. According to the second method, the tasks of a gang are candidate for migration only if at least one of them is at the head of its assigned queue and the respective processor is idle. The tasks that are migrated, are placed at the head of their newly assigned queues. In order to avoid the starvation of the other tasks, there is a limit on the number of migrated tasks a queue can accept.

Largest Gang First Served (LGFS)

According to this scheduling strategy, the tasks in the processor queues are sorted in descending order of gang size (i.e. number of tasks) of their respective gang. Thus, tasks that belong to larger gangs have higher priority than tasks that belong to smaller gangs. Whenever a processor becomes idle, the scheduler searches the queues starting from the head of each queue and the first gang with tasks that can start execution occupies the processors [19]. Clearly, this strategy tends to favor applications with a high degree of parallelism (i.e. large gangs), at the expense of smaller gangs. However, this is sometimes desirable and may lead to a better system performance, compared to the AFCFS policy.

3.2 *Coarse-Grained Parallel Applications*

In case an application exhibits *coarse-grained parallelism*, its component tasks do not require any communication with each other during processing, but only before or after their execution. That is, the component tasks have precedence constraints among them, in such a way that the output data of a task are used as input by other tasks. A component task can only start execution when its predecessor tasks have completed. A task without any parent tasks is called an *entry task*, whereas a task without any child tasks is called an *exit task*.

Such an application is often called a *workflow application* and can be represented by a *Directed Acyclic Graph (DAG)* or *task graph*, $G = (V, E)$, where V and E are the sets of the nodes and the edges of the graph, respectively [37, 39, 40]. Each node represents a component task, whereas a directed edge between two tasks represents the data that must be transmitted from the first task to the other. Each node has a weight that represents the computational cost of its corresponding task. Each edge between two tasks has a weight that denotes the communication cost that is incurred when transferring data from the first task to the other.

The *level* of a task in the graph is equal to the length of the longest path from the particular task to an exit task in the graph. The length of a path is the sum of the computational and communication costs of all of the nodes and edges, respectively, along the path. The *critical path* of the graph is the longest path from an entry task to an exit task in the graph. An example of a workflow application is illustrated in Fig. 4.

3.2.1 Workflow Scheduling Approaches

Workflow applications require a scheduling strategy that should take into account the precedence constraints among their component tasks. The workflow scheduling heuristics are classified into the following general categories:

- *list scheduling algorithms,*
- *clustering algorithms,*
- *task duplication algorithms* and
- *guided random search algorithms.*

These techniques are analyzed in the following paragraphs.

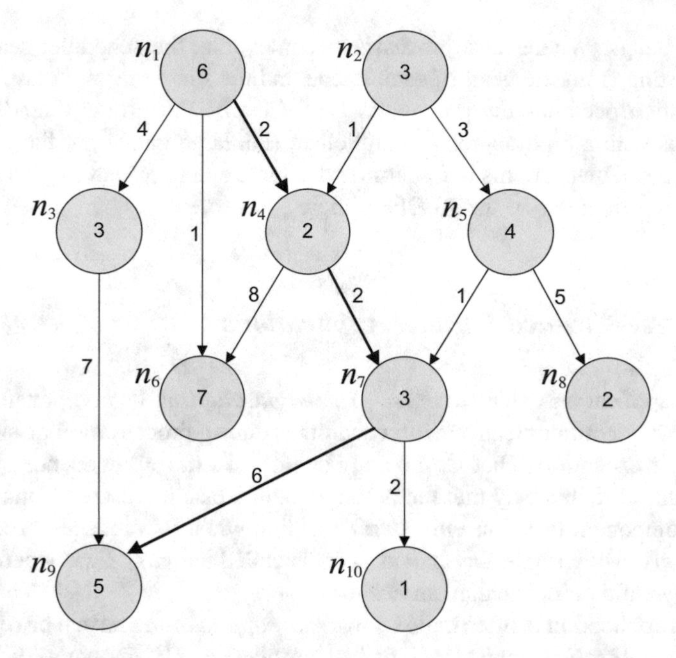

Fig. 4 An example of a coarse-grained parallel application (workflow application), represented as a Directed Acyclic Graph (DAG). The number in each node denotes the computational cost of the represented task. The number on each edge denotes the communication cost between the two tasks that it connects. The critical path of the DAG is depicted with thick arrows

List Scheduling Algorithms

A list scheduling algorithm consists of two phases: (a) a *task selection phase* and (b) a *processor selection phase*. In the first phase, the tasks are prioritized based on specific criteria and are arranged in a list according to their priority. The task with the highest priority is selected first for scheduling. During the second phase, the selected task is scheduled to the processor that minimizes a specific cost function, such as the estimated start time of the task [48]. List scheduling algorithms are the most commonly used among the workflow scheduling heuristics, because they are generally simpler, more practical, easier to implement and they usually outperform other techniques, incurring less scheduling overhead [53].

One of the simplest list scheduling policies is the *Highest Level First (HLF)* [1]. According to this method, the task prioritization phase is based on the level of each task. In the processor selection phase, the selected task is scheduled to the processor that can provide it with the earliest start time. An improved version of the HLF strategy is the *Insertion Scheduling Heuristic (ISH)* [21] and it is based on the observation that idle time slots may form in the schedule of a processor (schedule gaps), due to the data dependencies among the tasks. The task selection phase of this technique is based on HLF. However, during the processor selection phase, a task may be inserted into a schedule gap, as long as it does not delay the execution of the succeeding task in the schedule and provided that it cannot start earlier on any other processor. An alternative version of ISH, adapted for heterogeneous systems, is the *Heterogeneous Earliest Finish Time (HEFT)* policy [53]. According to this approach, for the calculation of the level of each task, the average computational and communication costs of the tasks and edges, respectively, are used.

Clustering Algorithms

The main idea of clustering algorithms is the minimization of the communication cost between the tasks of a DAG, by grouping heavily communicating tasks into the same cluster and assigning all of the tasks in the cluster to the same processor. A clustering algorithm is an iterative process. At first, each task is an independent cluster. At each iteration, previous clusters are refined by merging some of them, according to specific criteria. At the end of the process, a cluster merging step is needed, so that the number of clusters is equal to the number of processors. Subsequently, a cluster mapping step is required, in order to map each cluster to a processor. Finally, a task ordering step is performed, in order to determine the execution order of tasks on each processor [17].

One of the most popular clustering techniques is the *Dominant Sequence Clustering (DSC)* algorithm [57]. This method is based on the observation that the makespan of a DAG is determined by the longest path in the scheduled task graph and not by its critical path, which is calculated before the scheduling of the tasks of the DAG. The longest path in the scheduled DAG is called the *dominant sequence (DS)*. According to the DSC algorithm, the tasks in a DAG are clustered in such a way, so that the dominant sequence of the graph is minimized.

Task Duplication Algorithms

In this category of workflow scheduling heuristics, the main concept is to utilize idle resource time by duplicating predecessor tasks in a DAG, so that the makespan of the particular DAG is minimized. The various duplication-based algorithms differentiate with each other, according to the criteria used for the selection of the tasks for duplication. One of the major drawbacks of task duplication algorithms, is that they usually have higher complexity than the other DAG scheduling techniques.

One of the most well-known duplication algorithms is the *Duplication Scheduling Heuristic (DSH)* [21]. According to this approach, the tasks in a DAG are prioritized according to their level. At each scheduling step, the task with the highest level is selected and is allocated to the processor that can provide it with the earliest start time. In order to calculate the earliest possible start time of the selected task on each processor, first its start time is calculated without duplication of any predecessor tasks. Subsequently, the *duplication time slot* is determined, which is the time period between the finish time of the last scheduled task on the particular processor and the start time of the currently examined task. The algorithm then tries to duplicate the predecessors of the task into the duplication time slot in a recursive manner, starting from the parent task from which the data arrives the latest, until either the slot cannot accommodate other predecessor tasks or the start time of the examined task is not improved.

Guided Random Search Algorithms

A guided random search algorithm is an iterative process of finding the best schedule for a DAG, based on specific criteria. At each step, the previously generated schedule is improved, by utilizing random parameters for the generation of the new schedule. This iterative process terminates according to a predefined condition. These algorithms, even though they generally generate schedules of good quality, however, they incur a much higher scheduling overhead than the other workflow scheduling methods. The most commonly used algorithms of this category are *genetic algorithms*, according to which each new schedule is generated by applying evolutionary techniques from nature, known as *fitness functions* [15].

Simulated Annealing (SA) is another example of a guided random search metaheuristic. This technique emulates the physical process of annealing in metallurgy, which involves the heating and the controlled, slow cooling of metals, in order to form a crystallized structure without any defects [28]. In SA, a temperature variable is used in order to simulate this heating process. Initially, it is set at a high value and as the algorithm runs, it is allowed to slowly cool down. While the value of the temperature variable is high, the algorithm is allowed to accept solutions that are worse than the current one, with higher frequency. As the value of the temperature variable is decreased, so is the chance of accepting worse solutions. Therefore, the algorithm gradually focuses on an area of the search space in which hopefully a near-optimal solution can be found.

3.3 *Embarrassingly Parallel Applications*

An application is regarded as *embarrassingly parallel* when its component tasks are independent, do not communicate with each other and can be executed in any order. Due to these characteristics, such applications are also called *Bag-of-Tasks (BoT)* applications. Due to the independence between their tasks, BoT applications are well suited for execution on widely distributed resources, such as computational grids, where communication can become a bottleneck for more tightly-coupled parallel applications, such as gangs and DAGs [44, 46, 56]. An example of a BoT application is depicted in Fig. 5.

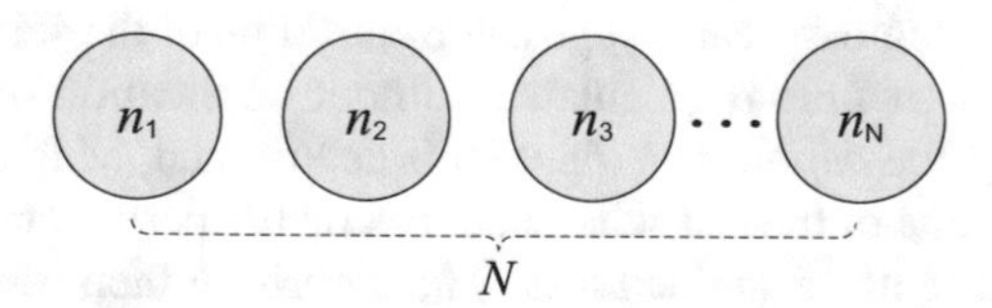

Fig. 5 An embarrassingly parallel application, consisting of N independent parallel tasks. Such applications are commonly referred to as Bag-of-Tasks (BoT) applications

3.3.1 Scheduling BoT Applications

The most widely used strategies for scheduling BoT applications are: (a) *Min-Min*, (b) *Max-Min* and (c) *Sufferage*. All of these policies focus on minimizing the makespan of the scheduled BoT application.

Min-Min

This heuristic is an iterative process, consisting of two steps. In the first step, the *minimum completion time (MCT)* of each unassigned task is calculated, over all of the processors in the system. In the second step, the task with the minimum MCT is assigned to the corresponding processor. At each iteration of the algorithm, the MCT of each unassigned task is determined taking into account the current load of the processors, as resulted by the scheduling decision of the previous iteration [56].

Max-Min

This strategy differs from the Min-Min policy, in that the task with the maximum (instead of the minimum) MCT is assigned to the corresponding processor in the second step of the scheduling process. Consequently, in cases where the application consists of a large number of tasks with small execution times and a few tasks with large execution times, the Max-Min heuristic is likely to give a smaller makespan than the Min-Min algorithm, since it schedules the tasks with larger execution times at earlier iterations [50].

Sufferage

This algorithm is a two-step iterative process, like the Min-Min and Max-Min heuristics. However, in this case, in addition to the MCT of each task, its second MCT is also calculated during the first step of the process. Subsequently, the *sufferage value* of each task is determined, by subtracting its MCT from its second MCT. In the second step, the task with the largest sufferage value is assigned to the processor that can provide it with the MCT. That is, this heuristic is based on the idea that the highest priority for scheduling should be given to the task that would suffer the most (in terms of completion time) if it is not assigned to the processor that can provide it with the MCT [24].

4 Major Challenges

In addition to the challenges imposed by their degree of parallelism, data-intensive applications in large-scale distributed systems must also effectively exploit data locality. Furthermore, they may have various QoS requirements, such as timeliness and fault tolerance, as well as other objectives, like energy efficiency. These requirements are usually specified in a *Service Level Agreement (SLA)*, which is a contract between the user that submits the application for execution and the provider of the infrastructure that the application is executed on. In the following paragraphs, representative examples for each case are given.

4.1 Data Locality

The most important aspect of scheduling data-intensive applications in large-scale distributed systems is the effective exploitation of data locality. That is, the tasks that operate on big data should be allocated to computational resources that are as near as possible to where the data reside, so that the communication cost incurred by transferring for processing vast amounts of data from remote resources is minimized.

4.1.1 MapReduce & Hadoop

The MapReduce programming paradigm has been proposed by Google [11] and facilitates the massively parallel processing of large volumes of data. It is inspired by the map and reduce functions commonly used in functional programming. A MapReduce application consists of two types of tasks: (a) a *map task* and (b) a *reduce task*. A map task takes a set of data and converts it into another set of data, where individual elements are broken down into tuples (i.e. key/value pairs). Parallel map tasks can process different chunks of data. A reduce task takes as input the output from map tasks and combines those data tuples into a smaller set of tuples, in order to produce the final result. A reduce task is always performed after the map tasks. In case a MapReduce application has only map tasks, it is considered an embarrassingly parallel application. In case an application has one or more reduce tasks, it is considered a coarse-grained parallel application. In the latter case, multiple reduce tasks can be employed in order to enhance the parallelism of the application [12].

A simple example of a MapReduce application with two parallel map tasks and one reduce task, is shown in Fig. 6. In the illustrated example, the overall minimum temperature recorded in London and Athens in a five-day period needs to be calculated for each city. It is assumed that the minimum temperature for each city was recorded daily in the form ⟨ *City*, *MinimumTemperature* ⟩. The records are split into two files. Each file is processed in parallel by a map task. Each map task outputs the pairs that correspond to the minimum temperature for each city, according to the file that was processed. The results of the two map tasks are merged into two pairs (one for each city) in the form ⟨ *City*, {*ListOfMinimumTemperatures*} ⟩. The pairs are fed as input into the reduce task, which outputs the overall minimum temperature recorded in each city, over the said period. This parallel processing approach is more efficient than calculating the minimum temperature for each city in a serial fashion.

An open source - and the most popular - implementation of the MapReduce programming model is the Apache Hadoop framework [2], which adopts a master-slave architecture in order to exploit data locality. Specifically, the master node is responsible for scheduling the map tasks of an application on the slave nodes, which contain chunks of the required input data. The reduce task is performed by the master node. When a slave node notifies the master node that it can accept a task, the master node scans the waiting tasks in queue to find the one that can achieve the best data locality. That is, the map task that its input data are located the nearest to the particular

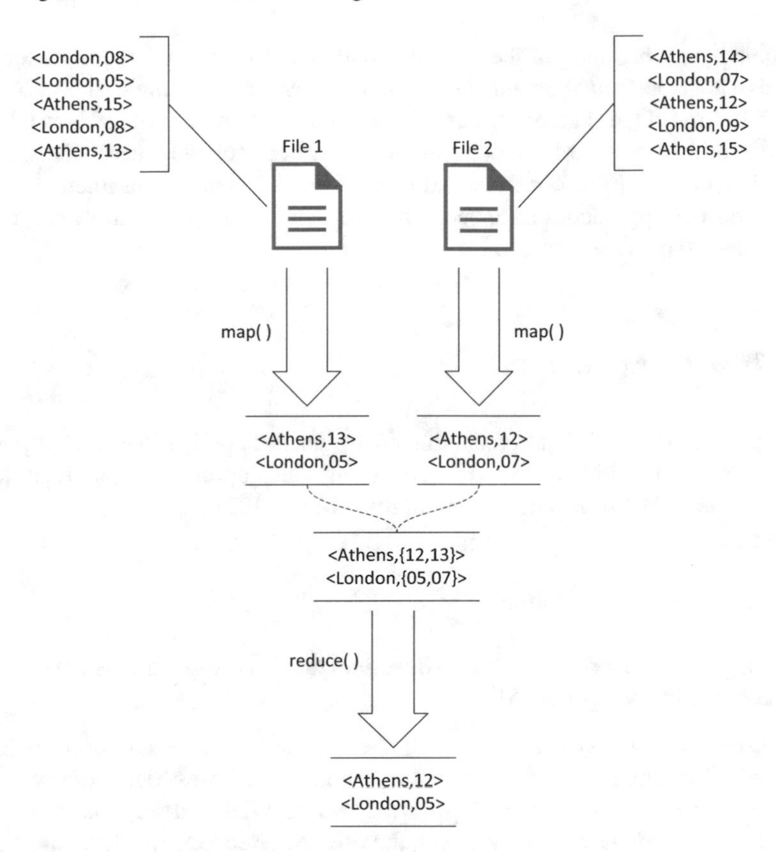

Fig. 6 An example of a MapReduce application with two parallel map tasks and one reduce task

slave node is selected. However, due to the fact that Hadoop considers only one slave node at a time in order to schedule the map tasks, there are cases where it does not exploit data locality effectively. Furthermore, it cannot be employed for multi-cluster processing and for data-intensive applications that require more complex communication and processing patterns than those supported by the MapReduce paradigm.

4.1.2 Other Approaches

In an attempt to tackle the aforementioned shortcomings of Hadoop and MapReduce, various approaches have been investigated in the literature. Among them, the *delay scheduling* technique has been proposed, in order to delay the scheduling of the waiting map tasks in case a slave node does not contain their input data, assuming that another slave node that contains the data will become available in a short period of time [58]. However, the drawback of this approach is that it wastes valuable time postponing the scheduling of the tasks, in an attempt to achieve better data locality, which is a goal that is not guaranteed. In order to overcome the single-cluster deployment restriction of the Hadoop framework, G-Hadoop has been proposed [55]. It is an extension of the traditional Hadoop framework that can schedule tasks across nodes of multiple clusters [59]. For the scheduling of more complex data-intensive applications, various approaches have been proposed, utilizing the workflow scheduling paradigms described in Sect. 3.2.1.

4.2 Time Constraints

The most common QoS requirement that data-intensive applications may impose, is to finish execution within a strict time constraint. Such applications are regarded as *real-time*, since they have a deadline that must be met [32].

4.2.1 Real-Time Applications

Depending on the severity of a missed deadline, real-time applications are classified into the following categories [5]:

- *Applications with soft deadlines*: in this case, the results of an application that missed its deadline still have some value, but their usefulness decreases with time (e.g. a user-system interaction application, where a delayed response to the user input is tolerated, degrading, however, the user experience as the delay increases).
- *Applications with firm deadlines*: in this case, the results will be useless, but this does not have any catastrophic consequences (e.g. a video streaming application, where a delayed video frame that arrives after the previous one on the user's terminal is discarded, since there is no value in playing it back).
- *Applications with hard deadlines*: in this case, not only will the results be useless, but missing the application's deadline will have catastrophic consequences. In this case, the damage caused by missing the deadline increases with time (e.g. a healthcare monitoring application, where a delayed analysis of patients data may cause loss of lives).

The impact of missing an application's deadline, as described above, is shown schematically in Fig. 7.

Two of the most widely used policies for the scheduling of real-time data-intensive applications are the *Earliest Deadline First (EDF)* and the *Least Laxity First (LLF)* algorithms [23, 27]. According to the EDF strategy, the component task with the highest priority for execution is the one with the earliest deadline. On the other hand, according to the LLF policy, the task with the highest priority is the one with the minimum *laxity*. The laxity of a task at a specific time instant, is defined as the difference between its deadline and its finish time. That is, it is the maximum amount of time that the particular task can delay its execution and still not miss its deadline.

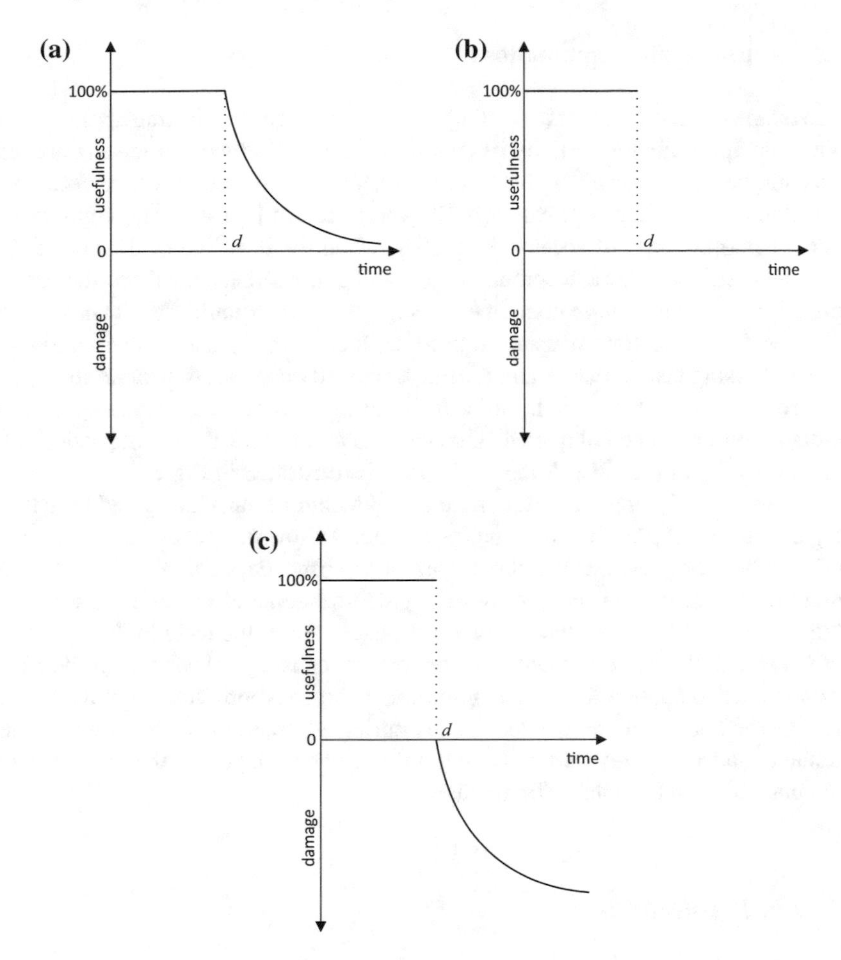

Fig. 7 The usefulness of the results of an application with a deadline d over time, when d is: **a** soft, **b** firm and **c** hard

A heuristic for the scheduling of real-time workflow applications in distributed systems, is the *Least Space-Time First (LSTF)* policy [8], which takes into account both the precedence and the time constraints among the tasks. Specifically, according to this method, the task with the highest priority for scheduling is the one with the minimum value of the *space-time* parameter. The space-time parameter of a task at a specific time instant, is defined as the difference between the deadline of the DAG and the level of the particular task. Even though this algorithm outperforms other scheduling policies, such as EDF, LLF and HLF described earlier, in the sense that it minimizes the maximum tardiness of the tasks, however, it exhibits poorer performance at guaranteeing deadlines.

4.2.2 Approximate Computations

Based on the observation that it is often more desirable for a real-time application to produce an approximate result by its deadline, than to produce a precise result late, the technique of *approximate computations* has been proposed [22]. According to this method, a real-time application is allowed to return intermediate, approximate results of poorer, but still acceptable quality, when the deadline of the application cannot be met. Approximate computations can be utilized especially in the case of applications with *monotone* component tasks, where the quality of a task's results is improved as more time is spent to produce them (e.g. statistical estimation and video processing tasks). Each monotone task typically consists of a *mandatory part*, followed by an *optional part*. In order for a task to return an acceptable result, its mandatory part must be completed. The optional part refines the result produced by the mandatory part [35, 36]. A monotone task is illustrated in Fig. 8.

Consequently, the approximate computations technique provides scheduling flexibility, by trading off precision for timeliness, since it allows the scheduler to terminate a task that has completed its mandatory part at any time, depending on the workload conditions of the system. For example, a video-on-demand server which streams video content to users over the Internet can benefit from this technique. The server may unexpectedly encounter network congestion, causing delays during the transmission of video content to users. Approximate computations can allow the system to reduce the quality of some video frames during a transmission, by omitting their optional enhancement layers and leaving only their base layer, so that the delivered video maintains an acceptable frame rate.

4.3 Fault Tolerance

Fault tolerant scheduling in large-scale distributed systems, such as clouds, is usually achieved through *application-directed checkpointing*, which in contrast to system-directed checkpointing, is more practical, easier to implement and system-independent [29]. According to this approach, each application is responsible for checkpointing its own progress periodically, at regular intervals during its execution.

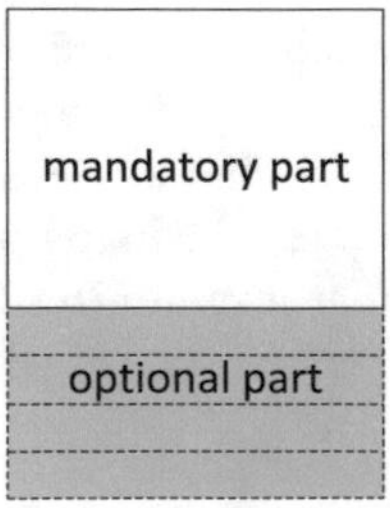

Fig. 8 A monotone task

mandatory part

optional part

In parallel data-intensive applications in particular, each component task periodically stores its state and intermediate data on persistent storage, creating a local checkpoint. The set of the local checkpoints (one from each task) that form a consistent application state, constitute a consistent global checkpoint.

When a failure occurs, the application is rolled back and resumes execution from its last consistent global checkpoint. Checkpointing is a reactive failure management technique, where recovery measures are taken after the occurrence of a failure. As opposed to proactive failure management approaches, where prevention measures are taken before the occurrence of a failure (e.g. task migrations), reactive management is simpler to implement, since it does not require any complex failure prediction methods.

4.4 Energy Efficiency

There is a growing focus on *green computing* from both the academia and the industry, in an attempt to minimize the carbon footprint of data centers and increase the energy efficiency of applications. Typically, in most computing systems the processor consumes the greatest amount of energy compared to other components [47, 54]. In embedded systems, as well as in large-scale virtualized platforms such as the cloud, a technique that is frequently used in order to meet the energy constraints is the *Dynamic Voltage and Frequency Scaling (DVFS)* method. This technique allows the dynamic adjustment of the supply voltage and operating frequency (i.e. speed) of a processor, based on the workload conditions, in an attempt to reduce the energy consumption of the processor [20, 52].

A heuristic frequently used with DVFS, is the *slack reclamation* technique [7]. This method is based on the fact that the actual execution time of tasks is sometimes much shorter than their estimated worst case execution time. The difference between the actual and the worst case execution time of a task is called *slack time*. At runtime, the scheduler tries to reclaim the slack time due to the early completion of a task, by selecting an unprocessed task to be executed at a slower processor speed via DVFS and thus save energy.

An energy-efficient scheduling strategy for real-time BoT applications in the cloud utilizing DVFS, is the *Cloud-Aware Energy-Efficient Scheduling (CAEES)* algorithm [6]. At each scheduling step, this method attempts to reduce the total energy consumption of the hosts, by selecting the most suitable virtual machine (VM) for the execution of each task, in an energy-wise manner. Specifically, the algorithm tries to schedule a task by examining specific criteria, starting from the best solution and gradually going to the worst solution: (a) the task is scheduled to a VM in use, without requiring an increase in its frequency, (b) the task is scheduled to a VM in use, but its operating frequency needs to be increased, (c) the task is scheduled to an idle VM, but there is at least one other VM on the same host that is not idle (i.e. the host is not idle) and (d) the task is scheduled to an idle VM on an idle host.

5 Recent Novel Ideas and Research Trends

In an attempt to provide even more effective scheduling solutions for data-intensive workloads in large-scale distributed systems, recent novel approaches have been proposed in the literature. As virtualization technologies evolve, a growing trend is the use of *VM live migrations*, in order to better exploit data locality. Another prominent research trend is the utilization of approximate computations in combination with other techniques, in order to achieve better scheduling performance, in terms of timeliness, resilience against failures and energy conservation. For example, approximate computations can be combined with:

- bin packing techniques, in order to enhance timeliness,
- checkpointing, in an attempt to improve fault tolerance and
- DVFS, for better energy efficiency.

5.1 VM Live Migrations

In virtualized platforms, the VM live migration technique refers to the process of moving a running VM from one physical host to another, without downtime. That is, with no impact on the availability of the VM to the end-users and without interrupting the applications currently running on the VM. The memory, storage and network connectivity of the VM are transferred from the initial physical host to the destination host. Currently, the predominant use of VM live migrations, is to enhance energy efficiency and load balancing through server consolidation [3].

However, the utilization of VM live migrations can also be used to better exploit data locality. Specifically, a virtualization approach has been proposed, where different VMs are used for each compute node and each storage node in the cloud [49]. In contrast to the traditional approach where each compute and storage node are combined into one VM, this approach provides better flexibility and scalability, since compute nodes and storage nodes can be added or removed from the cloud independently. More importantly, according to this approach, a much lower live migration cost is incurred by migrating a compute node VM, compared to the traditional approach, where large volumes of data should be transferred to the destination host, since a VM would be both a compute and a storage node. In this framework, a data-aware scheduling method, *DSFvH*, is employed, according to which live migrations of compute node VMs are performed, in order to place each compute node VM on the physical host that runs the storage node VM that contains the data required by the tasks executing on the compute node VM. This way, better exploitation of data locality is achieved.

5.2 Approximate Computations with Bin Packing

The traditional *bin packing* problem concerns the packing of a set of objects into a set of bins, using as few bins as possible [10]. The most commonly used bin packing techniques are: (a) *First Fit (FF)*, where the object is placed into the first bin where it fits, (b) *Best Fit (BF)*, where the object is placed into the bin where it fits and leaves the minimum unused space possible and (c) *Worst Fit (WF)*, where the object is placed into the bin where it fits and leaves the maximum unused space possible.

In an attempt to improve the timeliness of real-time workflow applications in a heterogeneous distributed system, a novel list scheduling heuristic has been proposed, which utilizes schedule gaps with a technique that combines approximate computations with the FF, BF and WF bin packing policies [38, 41]. Another characteristic of the proposed approach, is that it takes into account the effects of error propagation among the tasks of an application in case of partially completed tasks. The task prioritization is based on EDF. Once a task is selected by the scheduler, it is allocated to the processor that can provide it with the earliest estimated start time. In order to calculate the estimated start time of the task on the particular processor, schedule gaps are exploited with a technique that allows only a fraction of the task to be inserted into an idle time slot. The fraction of the task to be inserted into a schedule gap must be at least equal to the mandatory part of the task. Moreover, its potential output error must not exceed the input error limit of its child tasks.

The placement of the partial task into a schedule gap is performed using a modified version of either the FF, BF or WF bin packing policy:

- *First Fit with Approximate Computations (FF_AC)*: the task is placed into the first schedule gap where at least its minimum possible computational cost fits.
- *Best Fit with Approximate Computations (BF_AC)*: the task is placed into the schedule gap where its maximum possible computational cost fits, leaving the minimum unused time possible.
- *Worst Fit with Approximate Computations (WF_AC)*: the task is placed into the schedule gap where its minimum possible computational cost fits, leaving the maximum unused time possible.

In contrast to this approach, the other list scheduling heuristics presented earlier, ISH and HEFT, essentially use FF in order to utilize idle time slots. More importantly, with the incorporation of approximate computations, this approach is more flexible, allowing only a fraction of a task to be inserted into a schedule gap when the task does not completely fit into it. An example of scheduling tasks with the proposed heuristics (EDF_FF_AC, EDF_BF_AC and EDF_WF_AC), compared to the baseline EDF policy, is illustrated in Fig. 9. The parameters of the tasks used in the example are shown in Table 1.

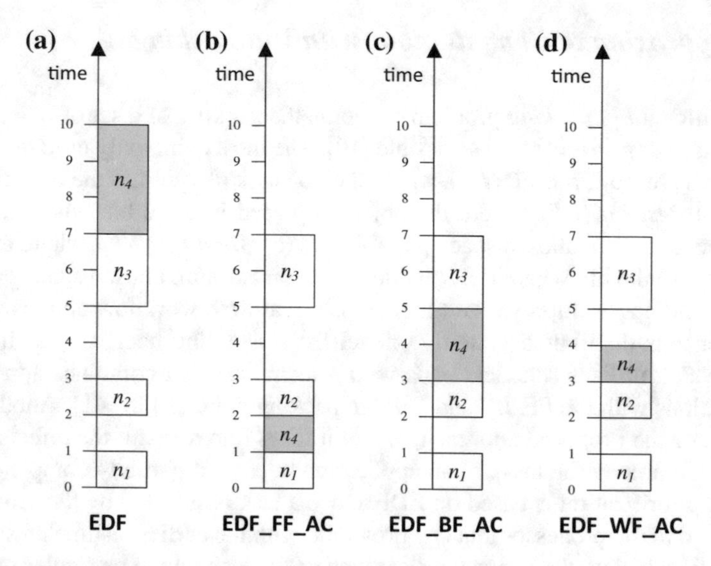

Fig. 9 An example of scheduling tasks with the strategies described in Sect. 5.2. A task n_4 is scheduled according to one of the policies: **a** EDF (baseline algorithm), **b** EDF_FF_AC, **c** EDF_BF_AC and **d** EDF_WF_AC. The parameters of the tasks used in the example are shown in Table 1

Table 1 The parameters of the tasks used in the example of Fig. 9. For each task, d is its deadline, t_{data} is the time at which its required input data will be available, c is its computational cost and c_{min} is its minimum computational cost when approximate computations are utilized

Task	d	t_{data}	c	c_{min}
n_1	2	0	1	1
n_2	4	2	1	1
n_3	9	5	2	1
n_4	10	1	3	1

5.3 *Approximate Computations with Checkpointing*

In an attempt to improve resilience against transient software failures in a SaaS cloud, where real-time fine-grained parallel applications are scheduled and executed, the approximate computations technique has been combined with application-directed checkpointing [33, 34, 43]. Specifically, gang scheduling is employed, where the prioritization of the component tasks is according to the EDF policy. In addition to application-directed checkpointing, fault tolerance is enhanced by the use of approximate computations in either a restricted manner or a more holistic approach. In the first case, an application may provide approximate results when it has completed its parallel mandatory part and (a) its deadline is reached, (b) a failure occurred and its last generated checkpoint stored results corresponding to computational work greater than or equal to its mandatory part or (c) another notified application must start execution immediately (i.e. there is time to execute only the mandatory part of the other application before its deadline). According to the second approach, all applications are scheduled to complete only their mandatory part. That is, in this case all applications give approximate results.

5.4 Approximate Computations with DVFS

In order to enhance energy efficiency, a heuristic that combines approximate computations with DVFS has been proposed, for the scheduling of periodic real-time tasks [26]. According to this approach, the tasks are scheduled according to the *Mandatory-First Earliest Deadline (MFED)* policy, while the supply voltage and processor frequency are scaled according to the *Cycle-Conserving Real-Time DVFS (CC-RT-DVFS)* technique. MFED is a policy according to which the mandatory parts of the tasks have always higher priority than the optional parts. The mandatory part with the earliest deadline has the highest priority for execution. CC-RT-DVFS is essentially a dynamic slack reclamation technique, which utilizes the slack time that occurs due to the early completion of a mandatory part, for the scheduling of the optional part of the task at a lower processor speed, utilizing DVFS. Thus, in this strategy there is a trade-off not only between result precision and timeliness, but also between result precision and energy savings.

6 Conclusions

In this chapter, a classification of data-intensive workloads was proposed and an overview of the most commonly used heuristics for their scheduling in large-scale distributed systems was given. Major challenges of data-intensive applications were covered, such as data locality awareness, timeliness, resilience against failures and energy efficiency. Furthermore, recent novel ideas and research trends were presented.

Scheduling data-intensive workloads in large-scale distributed systems remains an active research area, with many open challenges. With the explosive growth of big data, workloads tend to get more complex and computationally demanding. Consequently, more effective scheduling heuristics must be employed. In addition to the data locality awareness, timeliness, fault tolerance and energy efficiency objectives, security is drawing an ever-increasing interest from both the industry and the research community. Hence, efforts towards this direction are expected to be intensified in the near future.

Acknowledgements The second author of this chapter, Helen D. Karatza, has been invited as a trainer to the cHiPSet Training School 2016 "*New Trends in Modeling and Simulation in HPC Systems*", held in Bucharest, Romania, 21–23 September 2016, and has been supported by the IC1406 Horizon 2020 grant.

References

1. Adam, T.L., Chandy, K.M., Dickson, J.R.: A comparison of list schedules for parallel processing systems. Commun. ACM **17**(12), 685–690 (1974)
2. Apache: Apache Hadoop (2017). http://hadoop.apache.org/. Accessed 19 Jun 2017
3. Beloglazov, A., Abawajy, J., Buyya, R.: Energy-aware resource allocation heuristics for efficient management of data centers for cloud computing. Futur. Gener. Comput. Syst. **28**(5), 755–768 (2012)
4. Bonomi, F., Milito, R., Natarajan, P., Zhu, J.: Fog Computing: A Platform for Internet of Things and Analytics, pp. 169–186. Springer, Berlin (2014)
5. Buttazzo, G.C.: Hard Real-Time Computing Systems: Predictable Scheduling Algorithms and Applications, 3rd edn. Springer, Berlin (2011)
6. Calheiros, R.N., Buyya, R.: Energy-efficient scheduling of urgent bag-of-tasks applications in clouds through DVFS. In: Proceedings of the 6th IEEE International Conference on Cloud Computing Technology and Science (CloudCom'14), pp. 342–349 (2014)
7. Chen, J.J., Yang, C.Y., Kuo, T.W.: Slack reclamation for real-time task scheduling over dynamic voltage scaling multiprocessors. In: Proceedings of the 2006 IEEE International Conference on Sensor Networks, Ubiquitous and Trustworthy Computing (SUTC'06), pp. 358–365 (2006)
8. Cheng, B.C., Stoyenko, A.D., Marlowe, T.J., Baruah, S.K.: LSTF: a new scheduling policy for complex real-time tasks in multiple processor systems. Automatica **33**(5), 921–926 (1997)
9. Cisco: Fog computing and the internet of things: extend the cloud to where the things are. Technical Report C11-734435-00 04/15, San Jose, CA (2015)
10. Coffman Jr., E.G., Csirik, J., Galambos, G., Martello, S., Vigo, D.: Bin Packing Approximation Algorithms: Survey and Classification, pp. 455–531. Springer, Berlin (2013)
11. Dean, J., Ghemawat, S.: MapReduce: simplified data processing on large clusters. Commun. ACM **51**(1), 107–113 (2008)
12. Ekanayake, J., Fox, G.: High performance parallel computing with clouds and cloud technologies. In: Proceedings of the First International Conference on Cloud Computing (CloudComp'09), pp. 20–38 (2009)
13. Foster, I., Zhao, Y., Raicu, I., Lu, S.: Cloud computing and grid computing 360-degree compared. In: Proceedings of the 2008 Grid Computing Environments Workshop (GCE'08), pp. 1–10 (2008)
14. Garey, M.R., Johnson, D.S.: Computers and Intractability: A Guide to the Theory of NP-Completeness. W. H. Freeman and Company, New York (1979)
15. Gkoutioudi, K.Z., Karatza, H.D.: Multi-criteria job scheduling in grid using an accelerated genetic algorithm. J Grid Comput. **10**(2), 311–323 (2012)
16. Hashem, I.A.T., Yaqoob, I., Anuar, N.B., Mokhtar, S., Gani, A., Khan, S.U.: The rise of big data on cloud computing: review and open research issues. Inf. Syst. **47**, 98–115 (2015)
17. Jiang, H.J., Huang, K.C., Chang, H.Y., Gu, D.S., Shih, P.J.: Scheduling concurrent workflows in HPC cloud through exploiting schedule gaps. In: Proceedings of the 11th International Conference on Algorithms and Architectures for Parallel Processing (ICA3PP'11), pp. 282–293 (2011)
18. Karatza, H.D.: The impact of critical sporadic jobs on gang scheduling performance in distributed systems. Simul.: Trans. Soc. Model Simul. Int. **84**(2–3), 89–102 (2008)
19. Karatza, H.D.: Scheduling jobs with different characteristics in distributed systems. In: Proceedings of the 2014 International Conference on Computer, Information and Telecommunication Systems (CITS'14), pp. 1–5 (2014)

20. Kolodziej, J.: Evolutionary Hierarchical Multi-Criteria Metaheuristics for Scheduling in Large-Scale Grid Systems. Springer, Berlin (2012)
21. Kruatrachue, B., Lewis, T.G.: Duplication scheduling heuristic, a new precedence task scheduler for parallel systems. Technical Report. 87-60-3, Oregon State University, Corvallis, OR (1987)
22. Lin, K.J., Natarajan, S., Liu, J.W.S.: Imprecise results: utilizing partial computations in real-time systems. In: Proceedings of the 8th IEEE Real-Time Systems Symposium (RTSS'87), pp. 210–217 (1987)
23. Liu, C.L., Layland, J.W.: Scheduling algorithms for multiprogramming in a hard real-time environment. J. ACM **20**(1), 46–61 (1973)
24. Maheswaran, M., Ali, S., Siegel, H.J., Hensgen, D., Freund, R.F.: Dynamic mapping of a class of independent tasks onto heterogeneous computing systems. J. Parallel Distrib. Comput. **59**(2), 107–131 (1999)
25. Manickam, V., Aravind, A.: A fair and efficient gang scheduling algorithm for multicore processors. In: Proceedings of the 6th International Conference on Information Processing (ICIP'12), pp. 467–476 (2012)
26. Mizotani, K., Hatori, Y., Kumura, Y., Takasu, M., Chishiro, H., Yamasaki, N.: An integration of imprecise computation model and real-time voltage and frequency scaling. In: Proceedings of the 30th International Conference on Computers and Their Applications (CATA'15), pp. 63–70 (2015)
27. Mok, A.K.: Fundamental design problems of distributed systems for the hard real-time environment. PhD thesis, Massachusetts Institute of Technology, Cambridge, MA (1983)
28. Moschakis, I.A., Karatza, H.D.: Multi-criteria scheduling of bag-of-tasks applications on heterogeneous interlinked clouds with simulated annealing. J. Syst. Softw. **101**, 1–14 (2015)
29. Oldfield, R.A., Arunagiri, S., Teller, P.J., Seelam, S., Varela, M.R., Riesen, R., Roth, P.C.: Modeling the impact of checkpoints on next-generation systems. In: Proceedings of the 24th IEEE Conference on Mass Storage Systems and Technologies (MSST'07), pp. 30–46 (2007)
30. Papazachos, Z.C., Karatza, H.D.: Performance evaluation of gang scheduling in a two-cluster system with migrations. In: Proceeding 23rd IEEE International Parallel and Distributed Processing Symposium (IPDPS'09), pp. 1–8 (2009)
31. Russom, P.: Big data analytics. Technical Report TDWI Best Pract. Rep., Fourth Quart., TDWI Research (2011)
32. Stankovic, J.A., Spuri, M., Ramamritham, K., Buttazzo, G.C.: Deadline Scheduling for Real-Time Systems: EDF and Related Algorithms. Kluwer Academic Publishers, Dordrecht (1998)
33. Stavrinides, G.L., Karatza, H.D.: Performance evaluation of gang scheduling in distributed real-time systems with possible software faults. In: Proceedings of the 2008 International Symposium on Performance Evaluation of Computer and Telecommunication Systems (SPECTS'08), pp. 1–7 (2008)
34. Stavrinides, G.L., Karatza, H.D.: Fault-tolerant gang scheduling in distributed real-time systems utilizing imprecise computations. Simul.: Trans. Soc. Model Simul. Int. **85**(8), 525–536 (2009)
35. Stavrinides, G.L., Karatza, H.D.: Scheduling multiple task graphs with end-to-end deadlines in distributed real-time systems utilizing imprecise computations. J. Syst. Softw. **83**(6), 1004–1014 (2010)
36. Stavrinides, G.L., Karatza, H.D.: The impact of input error on the scheduling of task graphs with imprecise computations in heterogeneous distributed real-time systems. In: Proceedings of the 18th International Conference on Analytical and Stochastic Modeling Techniques and Applications (ASMTA'11), pp. 273–287 (2011)
37. Stavrinides, G.L., Karatza, H.D.: Scheduling multiple task graphs in heterogeneous distributed real-time systems by exploiting schedule holes with bin packing techniques. Simul. Model. Pract. Theor. **19**(1), 540–552 (2011)
38. Stavrinides, G.L., Karatza, H.D.: Scheduling real-time DAGs in heterogeneous clusters by combining imprecise computations and bin packing techniques for the exploitation of schedule holes. Futur. Gener. Comput. Syst. **28**(7), 977–988 (2012)

39. Stavrinides, G.L., Karatza, H.D.: The impact of resource heterogeneity on the timeliness of hard real-time complex jobs. In: Proceedings of the 7th International Conference on PErvasive Technologies Related to Assistive Environments (PETRA'14), Workshop on Distributed Sensor Systems for Assistive Environments (Di-Sensa), pp. 65:1–65:8 (2014)
40. Stavrinides, G.L., Karatza, H.D.: Scheduling real-time jobs in distributed systems-simulation and performance analysis. In: Proceedings of the 1st International Workshop on Sustainable Ultrascale Computing Systems (NESUS'14), pp. 13–18 (2014)
41. Stavrinides, G.L., Karatza, H.D.: A cost-effective and QoS-aware approach to scheduling real-time workflow applications in PaaS and SaaS clouds. In: Proceedings of the 3rd International Conference on Future Internet of Things and Cloud (FiCloud'15), pp. 231–239 (2015)
42. Stavrinides, G.L., Karatza, H.D.: Scheduling different types of applications in a saas cloud. In: Proceedings of the 6th International Symposium on Business Modeling and Software Design (BMSD'16), pp. 144–151 (2016)
43. Stavrinides, G.L., Karatza, H.D.: Scheduling real-time parallel applications in saas clouds in the presence of transient software failures. In: Proceedings of the 2016 International Symposium on Performance Evaluation of Computer and Telecommunication Systems (SPECTS'16), pp. 1–8 (2016)
44. Stavrinides, G.L., Karatza, H.D.: The effect of workload computational demand variability on the performance of a SaaS cloud with a multi-tier SLA. In: Proceedings of the IEEE 5th International Conference on Future Internet of Things and Cloud (FiCloud'17), pp. 10–17 (2017)
45. Stavrinides, G.L., Karatza, H.D.: Periodic scheduling of mixed workload in distributed systems. In: Proceedings of the 23rd ICE/IEEE International Conference on Engineering, Technology and Innovation (ICE'17) (2017, in press)
46. Stavrinides, G.L., Karatza, H.D.: Scheduling real-time bag-of-tasks applications with approximate computations in SaaS clouds. Concurr. Comput. Pract. Exp. (2017, in press)
47. Stavrinides, G.L., Karatza, H.D.: Simulation-based performance evaluation of an energy-aware heuristic for the scheduling of HPC applications in large-scale distributed systems. In: Proceedings of the 8th ACM/SPEC International Conference on Performance Engineering (ICPE'17), 3rd International Workshop on Energy-aware Simulation (ENERGY-SIM'17), pp. 49–54 (2017)
48. Stavrinides, G.L., Duro, F.R., Karatza, H.D., Blas, J.G., Carretero, J.: Different aspects of workflow scheduling in large-scale distributed systems. Simul. Model. Pract. Theor. **70**, 120–134 (2017)
49. Sun, R., Yang, J., Gao, Z., He, Z.: A virtual machine based task scheduling approach to improving data locality for virtualized hadoop. In: Proceedings of the 2014 IEEE/ACIS 13th International Conference on Computer and Information Science (ICIS'14), pp. 297–302 (2014)
50. Tabak, E.K., Cambazoglu, B.B., Aykanat, C.: Improving the performance of independent task assignment heuristics minmin, maxmin and sufferage. IEEE Trans. Parallel. Distrib. Syst. **25**(5), 1244–1256 (2014)
51. Talia, D.: Clouds for scalable big data analytics. Computer **46**(5), 98–101 (2013)
52. Terzopoulos, G., Karatza, H.D.: Bag-of-task scheduling on power-aware clusters using a DVFS-based mechanism. In: Proceedings of the 28th IEEE International Parallel & Distributed Processing Symposium (IPDPS'14), 10th Workshop on High-Performance, Power-Aware Computing (HPPAC'14), pp. 833–840 (2014)
53. Topcuoglu, H., Hariri, S., Wu, M.Y.: Performance-effective and low-complexity task scheduling for heterogeneous computing. IEEE Trans. Parallel. Distrib. Syst. **13**(3), 260–274 (2002)
54. Valentini, G.L., Lassonde, W., Khan, S.U., Allah, N.M., Madani, S.A., Li, J., Zhang, L., Wang, L., Ghani, N., Kolodziej, J., Li, H., Zomaya, A.Y., Xu, C.Z., Balaji, P., Vishnu, A., Pinel, F., Pecero, J.E., Kliazovich, D., Bouvry, P.: An overview of energy efficiency techniques in cluster computing systems. Clust. Comput. **16**(1), 3–15 (2013)
55. Wang, L., Tao, J., Ranjan, R., Marten, H., Streit, A., Chen, J., Chen, D.: G-Hadoop: MapReduce across distributed data centers for data-intensive computing. Futur. Gener. Comput. Syst. **29**(3), 739–750 (2013)

56. Weng, C., Lu, X.: Heuristic scheduling for bag-of-tasks applications in combination with QoS in the computational grid. Futur. Gener. Comput. Syst. **21**(2), 271–280 (2005)
57. Yang, T., Gerasoulis, A.: DSC: scheduling parallel tasks on an unbounded number of processors. IEEE Trans. Parallel. Distrib. Syst. **5**(9), 951–967 (1994)
58. Zaharia, M., Borthakur, D., Sen Sarma, J., Elmeleegy, K., Shenker, S., Stoica, I.: Delay scheduling: a simple technique for achieving locality and fairness in cluster scheduling. In: Proceedings of the 5th European Conference on Computer Systems (EuroSys'10), pp. 265–278 (2010)
59. Zhao, J., Wang, L., Tao, J., Chen, J., Sun, W., Ranjan, R., Kolodziej, J., Streit, A., Georgakopoulos, D.: A security framework in G-Hadoop for big data computing across distributed cloud data centres. J. Comp. Syst. Sci. **80**(5), 994–1007 (2014)

Design Patterns and Algorithmic Skeletons: A Brief Concordance

Adriana E. Chis and Horacio González–Vélez

Abstract Having been designed as abstractions of common themes in object-oriented programming, patterns have been incorporated into parallel programming to allow an application programmer the freedom to generate parallel codes by parameterising a framework and adding the sequential parts. On the one hand, parallel programming patterns and their derived languages have maintained, arguably, the best adoption rate; however, they have become conglomerates of generic attributes for specific purposes, oriented towards code generation rather than the abstraction of structural attributes. On the other hand, algorithmic skeletons systematically abstract commonly-used structures of parallel computation, communication, and interaction. Although there are significant examples of relevant applications—mostly in academia—where they have been successfully deployed in an elegant manner, algorithmic skeletons have not been widely adopted as patterns have. However, the ICT industry expects graduates to be able to easily adapt to its best practices. Arguably, this entails the use of pattern-based programming, as it has been the case in sequential programming where the use of design patterns is widely considered the norm, as demonstrated by a myriad of citations to the seminal work of Gamma et al. [6] widely known as the Gang-of-Four. We contend that an algorithmic skeleton can be treated as a structural design pattern where the degree of parallelism and computational infrastructure are only defined at runtime. The purpose of this chapter is to explain how design patterns can be mapped into algorithmic skeletons. We illustrate our approach using a simple example using the visitor design pattern and the task farm algorithmic skeleton.

Adriana E. Chis (✉) · H. González–Vélez
Cloud Competency Centre, National College of Ireland, Dublin 1, Ireland
e-mail: adriana.chis@ncirl.ie

H. González–Vélez
e-mail: horacio@ncirl.ie

J. Kołodziej et al. (eds.), *Modeling and Simulation in HPC and Cloud Systems*,
Studies in Big Data 36, https://doi.org/10.1007/978-3-319-73767-6_3

1 Introduction

Parallel programming aims to capitalise on concurrency, the execution of different sections of a given program at the same time, in order to improve the overall performance of the program, and, eventually, that of the whole system. Despite major breakthroughs, parallel programming is still a highly demanding activity widely acknowledged to be more difficult than its sequential counterpart, and one for which the use of efficient programming models and structures has long been sought. These programming models must necessarily be performance-oriented, and are expected to be defined in a scalable structured fashion to provide guidance on the execution of their jobs and assist in the deployment of heterogeneous resources and policies.

Furthermore, it is widely acknowledged that one of the major challenges of the multi/many-core era is the efficient support of parallel programming models that can predict and improve performance for diverse heterogenous architectures [11]. Furthermore, the "Berkeley View" work established the importance of not only producing realistic benchmarks for parallel programming models based on patterns of computation and communication, but also developing programming paradigms which efficiently deploy scalable task parallelism [2]. Such decoupling has allowed them to be efficiently deployed on different dedicated and non-dedicated architectures including symmetric multiprocessing, massively parallel processing, clusters, constellations, and clouds.

Design patterns have been conceived as abstractions of common themes in object-oriented programming [5, 6]. *Parallel patterns* aim to further expand this concept by decoupling the detail or implementation from the structure of a parallel program in order to transfer any performance improvements in the system infrastructure while preserving the final result.

Algorithmic skeletons abstract commonly-used patterns of parallel computation, communication and interaction [3]. Skeletons provide a clear and consistent behaviour across platforms, with the underlying structure depending on the particular implementation [7].

Diverse authors have established the importance of patterns and skeletons in parallel programming from a design point of view [8–10, 12], and their benefits for applicative environments and development projects.

In this work, we give an initial example for a direct mapping of design patterns and skeletons in order to establish a correspondence between the well-known, accepted design pattern approach and the programmer-oriented functional algorithmic skeleton paradigm. This is not intended as a comprehensive survey but rather an initial attempt to introduce the topic to early career researchers and practitioners.

This chapter is structured as follows. Firstly, Sect. 2 provides a brief introduction to design patterns. Secondly, Sect. 3 describes the algorithmic skeleton paradigm. Thirdly, Sect. 4 describes our mapping of a design pattern to an algorithmic skeleton. Finally, Sect. 5 presents our conclusions.

2 Design Patterns

Computers have been traditionally programmed with a sequential frame of mind, but parallel solutions require a different way of approaching and dissecting a problem. They require a holistic analysis and understanding of the system architecture, the programming paradigm, and the problem constraints. Parallel computing requires calculations to be synchronised, staged, and/or communicated over a number of different phases. Message-passing, threads, load-balancing, and semaphores are matters restricted to the expert software developers and, arguably, lack some high-level design features required for large-scale software development endeavours.

Having defined a *pattern* as a core solution to a problem that recurrently occurs in a given context, Alexander et al. introduced a pattern language to describe tens of patterns applied in civil engineering [1]. Subsequently, design patterns have documented solutions to recurrent software design problems. Gamma et al. present 23 design patterns [6].

The authors classify the design patterns based on their purpose into three main categories, namely *creational patterns*, *structural patterns* and *behavioural patterns* as illustrated in Fig. 1.

Creational patterns	Used to build objects such that they can be decoupled from the implementing system.
Structural patterns	Used to form large data structures from many disparate objects.
Behavioural patterns	Used to manage algorithms, relationships, and responsibilities between objects.

Furthermore, Gamma et al. provide another classification of the design patterns based on the patterns' scope in *object patterns* and *class patterns*. As the name suggests the former category of patterns specify relationships between objects, whereas the latter category of patterns encodes relationships between classes and subclasses.

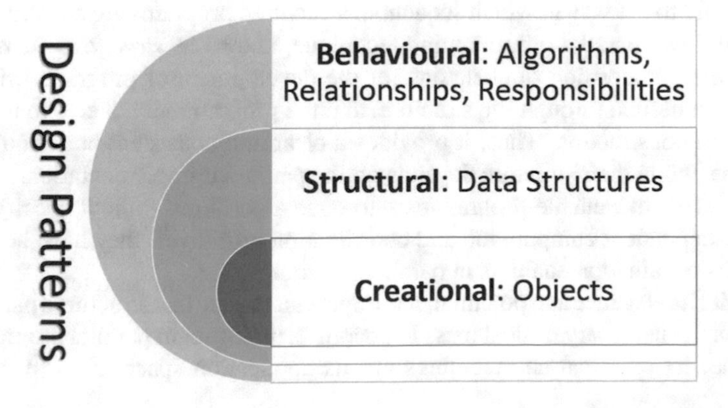

Fig. 1 Traditional classification of design patterns: creational, structural and behavioural

Table 1 Examples of design patterns (creational, structural, and behavioural) (*Source* [6])

Category	Pattern
Creational	Abstract Factory, Builder, Factory Method, Prototype, Singleton
Structural	Adapter, Bridge, Composite, Decorator, Facade, Flyweight, Proxy
Behavioural	Chain of Responsibility, Command, Interpreter, Iterator, Mediator, Memento, Observer, State, Strategy, Template Method, Visitor

A complete description of these design patterns can be found in the seminal book by Gamma et al. [6]. The authors document each pattern using a template. The core elements of describing a pattern are: the pattern name; the problem, which presents details about a problem in a given context; the solution in form of a generic design solution which incorporates the relationships and interactions between objects and classes; and the consequences of using a given pattern. A number of core design patterns are presented in Table 1.

3 Algorithmic Skeletons

Cole pioneered the field with the definition of skeletons as "specialised higher-order functions from which one must be selected as the outermost purpose in the program", and the introduction of four initial skeletons: divide and conquer, iterative combination, cluster, and task queue [3]. His work described a software engineering approach to high-level parallel programming using a skeletal (virtual) machine rather than the deployment of a tool or language on a certain architecture.

In essence, algorithmic skeletons systematically abstract commonly-used structures of parallel computation, communication, and interaction. Skeletal parallel programs are typically expressed by interweaving parameterised skeletons using descending composition and control inheritance throughout the program structure, analogously to the way in which sequential structured programs are constructed [4]. This high-level parallel programming technique, known as *structured parallelism*, enables the composition of skeletons for the development of programs where the control is inherited through the structure, and the programmer adheres to top-down design and construction. Thus, it provides a clear and consistent behaviour across platforms, while their structure depends on the particular implementation.

Since skeletons enable programmers to code algorithms without specifying the machine-dependent computation and coordination primitives, they have been positioned as coordination enablers in parallel programs.

Despite its elegance and potential, it is important to state that structured parallelism still lacks the necessary critical mass to become a mainstream parallel programming technique. Its principal shortcomings are its application space, since it can only

Table 2 A taxonomy for the algorithmic skeleton constructs based on their functionality

Skeleton	Scope	Main coordination characteristic	Examples
Data-parallel	Data structures	I/O intensive	Map, reduce
Task-parallel	Tasks	Scheduling	Task farm, pipeline
Resolution	Family of problems	Computational-intensive	Divide-and-conquer, branch-and-bound

address well-defined algorithmic solutions, and the lack of a specification to define and exchange skeletons between different implementations.

Skeletons can be broadly categorised into three types based on their functionality as shown in Table 2.

Data-parallel skeletons — Work typically on bulk data structures. Their behaviour establishes functional correspondences between data, and their structure regulates resource layout at fine-grain parallelism, e.g. MPI collectives.

Task-parallel skeletons — Operate on tasks. Their behaviour is determined by the interaction between tasks, and their coarse-grain structure establishes scheduling constraints among processes, e.g. task farm and pipeline.

Resolution skeletons — Delineate an algorithmic method to undertake a given family of problems. Their behaviour reflects the nature of the solution to a family of problems, and their structure may encompass different computation, communication, and control primitives, e.g. the divide-and-conquer and dynamic programming skeletons.

From a coordination point of view, data-parallel skeletons are typically input/output intensive as they operate on memory, and even disk, stored data structures, while resolution are computational-intensive as they deploy complex algorithms with demanding computational requirements. Task parallel can be construed as task schedulers with intimate knowledge of the program structure.

3.1 A Classification for Algorithmic Skeletons

This section elaborates on the functionality associated with the specific algorithmic skeletons listed in Table 2.

- Data-parallel (see Fig. 2)
  - *Map* specifies that a function or a sub-skeleton can be applied simultaneously to all the elements of a list to achieve parallelism. The data parallelism occurs

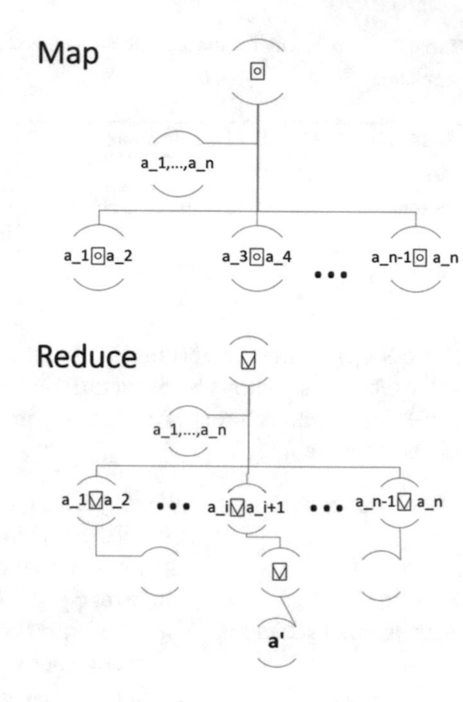

Fig. 2 Two data parallel skeletons: map and reduce

because a single data element can be split into multiple data, then the sub-skeleton is executed on each data element, and finally the results are united again into a single result. The *map* skeleton can be conceived as single instruction, multiple data parallelism.
  - *Reduce*, also known as *scan*, is employed to compute prefix operations in a list by traversing the list from left to right and then applying a function to each pair of elements, typically summation. Note that as opposed to *map*, it maintains aggregated partial results.

- Task-parallel (see Fig. 3)
  - *Task Farm* or simply *farm* embeds the ability to schedule independent tasks in a divisible workload across multiple computing nodes.
  - *Pipe* enables staged computations, where parallelism can be achieved by computing different stages simultaneously on different inputs. The number of stages provided by *pipe* can be variable or fixed.
- Resolution (see Fig. 4)
  - *Divide & Conquer* (d&c) calls are recursively applied until a condition is met within an optimisation search space. Its semantics are as follows. When an input arrives, a condition component is invoked on the input. Depending on the result two things can happen. Either the parameter is passed on to the sub-skeleton,

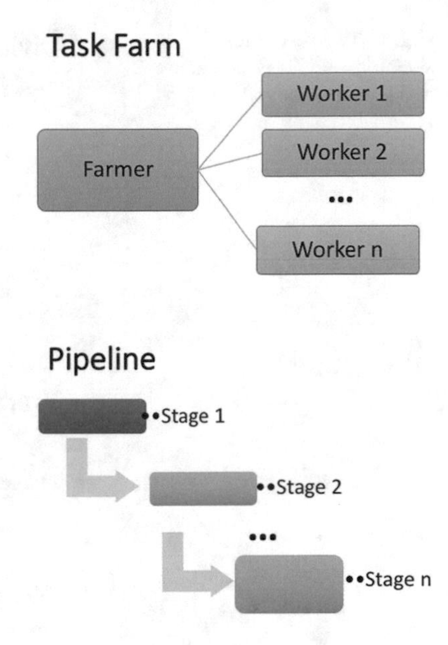

Fig. 3 Two task parallel skeletons: task farm and pipeline

or the input is split with the split component into a list of data. Then, for each list element the same process is applied recursively. When no further recursions are performed, the results obtained at each level are merged. Eventually, the merged results yield to one result which corresponds to the final result of the d&c skeleton.

- *Branch & Bound* (b&b) divides recursively the search space (branch) and then determines the elements in the resulting sub-spaces by mapping an objective function (bound). The merged results also produce one result which corresponds to the final result of the b&b skeleton.

4 Mapping Patterns and Skeletons

In this section we show how the *Visitor* pattern, a behavioural design pattern, can be mapped to the *Task Farm* algorithmic skeleton by documenting the latter using the design pattern template and identifying the commonalities between the two.

Design patterns comprise intent, motivation, participants, collaborations, and consequences. Given that structural design patterns have been conceived to create added functionality via object augmentation, they can be "made" parallel. That is to say, a standard compound structural pattern can have parallel characteristics which can be instantiated dynamically. On the other hand, algorithmic skeletons operate on

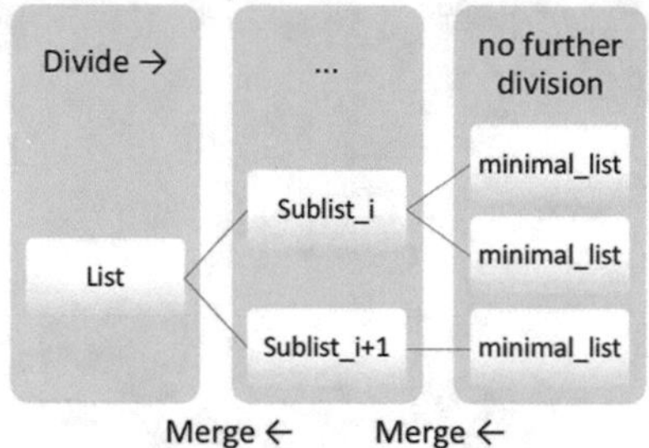

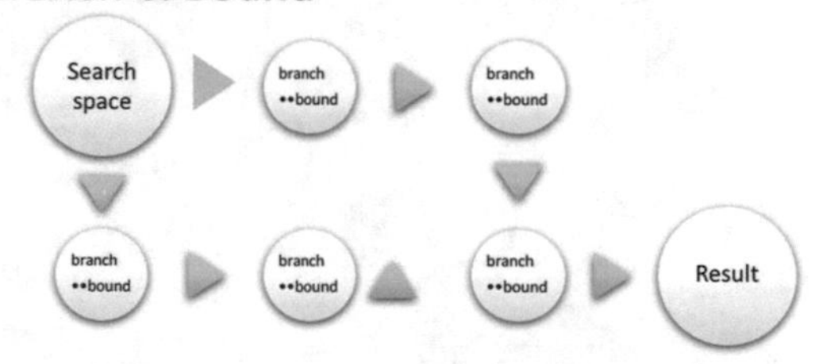

Fig. 4 Two resolution skeletons: divide & conquer and branch & bound

the notion of changing underlying computational resources and therefore detach the structure from the behaviour of the program.

We contend that an algorithmic skeleton can be treated as a structural design pattern where the degree of parallelism and computational infrastructure are only defined at runtime. From this perspective, the programmer task is arguably simplified by completely detaching the structure and behaviour as originally intended, and additionally increasing its consistency and programmability through the design pattern characteristics.

Visitor Pattern—A Behavioural Pattern

As introduced by Gamma et al. [6], the *Visitor* pattern separates structure from computation by enabling new operations on existing object structures without modifying the structures. Figure 5 presents the generic solution of the Visitor pattern. The following description is adapted from [6]:

Intent	Represent an operation to be performed on the elements of an object structure. Visitor lets you define a new operation without changing the classes of the elements on which it operates.
Motivation	Represent an operation to be performed on the elements of an object structure. Visitor lets you define a new operation without changing the classes of the elements on which it operates.
Applicability	The visitor pattern is useful in the following scenarios:

- An object structure contains many classes of objects with differing interfaces, and you want to perform operations on these objects that depend on their concrete classes.

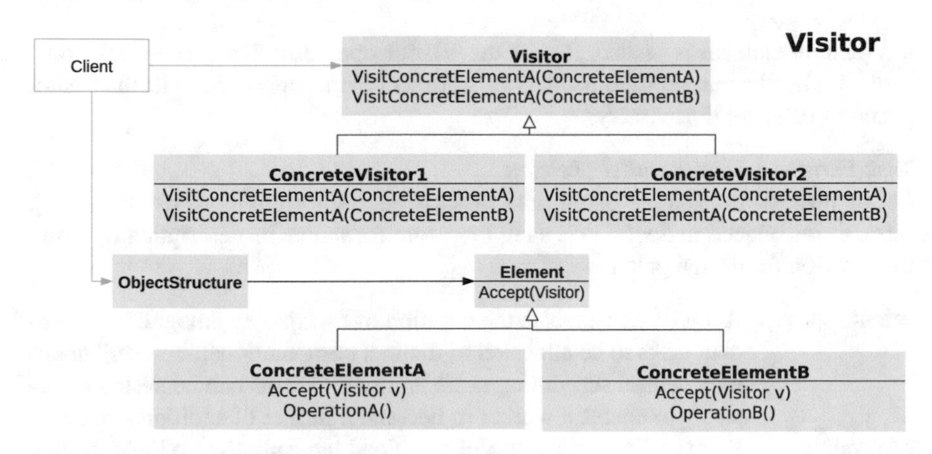

Fig. 5 Visitor pattern (*source* [6])

- Many distinct and unrelated operations need to be performed on objects in an object structure, and you want to avoid "polluting" their classes with these operations. Visitor lets you keep related operations together by defining them in one class. When the object structure is shared by many applications, use Visitor to put operations in just those applications that need them.
- The classes defining the object structure rarely change, but you often want to define new operations over the structure. Changing the object structure classes requires redefining the interface to all visitors, which is potentially costly. If the object structure classes change often, then it is probably better to define the operations in those classes.

Participants (consult the pattern structure shown in Fig. 5)

Visitor declares a *visit* method for each of the concrete elements that need to be traversed.

ConcreteVisitor implements each *visit* method declared in the Visitor. Usually, each ConcreteVisitor keeps track of a local state for the visited concrete element. The state is going to be updated while recursively traversing the structure.

Element declares an *accept* method which allows passing in a Visitor as a parameter

ConcreteElement declares an *accept* method which allows passing in a Visitor as a parameter

ObjectStructure offer a mechanism to allow a visitor to visit the elements

Collaborations

- A client that uses the Visitor pattern must create a *ConcreteVisitor* object and then traverse the object structure, visiting each element with the visitor.

- When an element is visited, it calls the Visitor operation that corresponds to its class. The element supplies itself as an argument to this operation to let the Visitor access its state, if necessary.

Task Farm—A Task-Parallel Skeleton

Let us present an example based on a task parallel algorithmic skeleton, the task farm, as introduced in Sect. 3. We shall therefore formalise its description by using the notation for design patterns.

Intent	A Task Farm enables the creation of a variable number of independent tasks to be allocated to distinct computational "worker" nodes by a central scheduling node "farmer". Farms can be nested recursively to enable a worker to become a farmer of additional nodes.
Motivation	Farms are especially useful to offload large numbers of independent tasks to several nodes. Typically there are many more tasks than nodes. As nodes can have different architectures (e.g. based on CPUs or GPUs) and, consequently, distinct computational characteristics, the farmer requires to allocate tasks using greedy or other heuristics scheduling mechanisms. Furthermore, computational resources may not necessarily be dedicated, can be geographically distributed, and have variable latencies, making the overall scheduling dynamic and complex.
Applicability	Farms are particularly useful to offload embarrassingly-parallel computations where the ordering and finish times of independent tasks are not subject to hard constraints.

Participants

Farmer the process which divides and allocates tasks to workers.
Worker the processes which receive tasks and compute results based on given function.

Collaborations

- A client that uses the Task Farm skeleton must create a Farmer object to create an object structure for Workers. By traversing the object structure for each Worker element, a Farmer assigns tasks to each Worker.
- When an element (Worker) completes a task (or a series of them), it calls the Farmer operation that corresponds to its class.

We notice that if we perform a pairwise comparison that the visitor design pattern and the task farm skeleton can be similar in nature. For instance, the intent of both is to perform a series of tasks without altering the nature of the structure and both have an architecture-independent approach. While the Task Farm deals with processing heterogeneity by using scheduling mechanisms, the Visitor pattern makes no assumption on the nature of the underlying infrastructure.

5 Conclusions

This initial mapping approach of patterns to skeletons has shown that, in principle, parallel programming structures can be formally documented using a design pattern notation to strengthen its nature and, most certainly, its readability.

With respect to the analysis of the mapping problem, the findings of this work provide an initial idea to document large parallel programming endeavours. This tacitly reinforces the notion that although parallel programming is complex, well-documented parallel patterns can help to ease the burden.

From a performance standpoint, it is arguable that the overall performance of algorithmic skeletons can be assumed to be unaltered as the design pattern notation is mostly static. Furthermore, by assuming a skeleton is a pattern whose degree of parallelism is determined at run-time, there is an intrinsic reinforcement to the decoupling of computation from coordination.

However, it is a fact there are substantial avenues of research that need to be explored to fully formalise a design pattern approach to skeletons.

Acknowledgements The authors would like to acknowledge the contribution of the ICT COST Action IC1406 "High-Performance Modelling and Simulation for Big Data Applications (cHiPSet)" http://chipset-cost.eu/.

References

1. Alexander, C., Ishikawa, S., Silverstein, M.: A Pattern Language: Towns, Buildings, Construction. Oxford University Press, New York (1977)
2. Asanovic, K., Bodik, R., Demmel, J., Keaveny, T., Keutzer, K., Kubiatowicz, J., Morgan, N., Patterson, D., Sen, K., Wawrzynek, J., Wessel, D., Yelick, K.: A view of the parallel computing landscape. Commun. ACM **52**(10), 56–67 (2009)
3. Cole, M.: Algorithmic Skeletons: Structured Management of Parallel Computation, Research Monographs in Parallel and Distributed Computing. MIT Press/Pitman, London (1989)
4. Cole, M.: Bringing skeletons out of the closet: a pragmatic manifesto for skeletal parallel programming. Parallel Comput. **30**(3), 389–406 (2004)
5. Gamma, E., Helm, R., Johnson, R., Vlissides, J.: Design patterns: Abstraction and reuse of object-oriented design. ECOOP'93. Lecture Notes in Computer Science, vol. 707, pp. 406–431. Springer, Kaiserslautern (1993)
6. Gamma, E., Helm, R., Johnson, R., Vlissides, J.: Design Patterns: Elements of Reusable Object-oriented Software. Addison-Wesley Longman, Boston (1995)
7. González-Vélez, H., Leyton, M.: A survey of algorithmic skeleton frameworks: high-level structured parallel programming enablers. Softw., Pract. Exper. **40**(12), 1135–1160 (2010)
8. Goswami, D., Singh, A., Preiss, B.R.: From design patterns to parallel architectural skeletons. J. Parallel Distrib. Comput. **62**(4), 669–695 (2002)
9. Hammond, K., Aldinucci, M., Brown, C., Cesarini, F., Danelutto, M., González-Vélez, H., Kilpatrick, P., Keller, R., Rossbory, M., Shainer, G.: The ParaPhrase Project: Parallel Patterns for Adaptive Heterogeneous Multicore Systems, Lecture Notes in Computer Science, vol. 7542, pp. 218–236. Springer, Torino (2013)
10. Mattson, T.G., Sanders, B., Massingill, B.: Patterns for Parallel Programming. Software Patterns Series. Addison-Wesley Professional, Boston (2004)

11. Mittal, S., Vetter, J.S.: A survey of CPU-GPU heterogeneous computing techniques. ACM Comput. Surv. **47**(4), 69:1–69:35 (2015)
12. Rabhi, F.A., Gorlatch, S. (eds.): Patterns and Skeletons for Parallel and Distributed Computing. Springer, London (2003)

Evaluation of Cloud Systems

Mihaela-Andreea Vasile, George-Valentin Iordache, Alexandru Tudorica and Florin Pop

Abstract Modelling and simulation represent suitable instruments for evaluation of distributed system. These essential tools in science are used in Cloud systems design and performance evaluation. The chapter covers the fundamental skills for a practitioner working in the field of Cloud Systems to have, for the development of a correct methodology for the evaluation using simulation of Cloud services and components. We concentrate on subjects related to tasks scheduling and resource allocation with the focus on scalability and elasticity, the constraints imposed by SLA and the use of CloudSim for performance evaluation of Cloud Systems. Several metrics used in modelling and simulation are presented in this chapter.

1 Introduction

Cloud services are classified by NIST into three categories [1]: *IaaS*—Infrastructure as a Service, *PaaS*—Platform as a Service, *SaaS*—Software as a Service.

This work was presented during the event cHiPSet Training School "*New Trends in Modeling and Simulation in HPC Systems*" Bucharest, Romania, 21–23 September 2016, supported by cHiPSet ICT COST Action IC1406.

M.-A. Vasile · G.-V. Iordache · A. Tudorica · F. Pop (✉)
Computer Science Department, Faculty of Automatic Control and Computers,
University Politehnica of Bucharest, Bucharest, Romania
e-mail: florin.pop@cs.pub.ro

M.-A. Vasile
e-mail: mihaela.vasile@hpc.pub.ro

G.-V. Iordache
e-mail: george.iordache@cs.pub.ro

A. Tudorica
e-mail: alexandru.tudorica@cti.pub.ro

F. Pop
National Institute for Research and Development in Informatics (ICI),
Bucharest, Romania

J. Kołodziej et al. (eds.), *Modeling and Simulation in HPC and Cloud Systems*,
Studies in Big Data 36, https://doi.org/10.1007/978-3-319-73767-6_4

IaaS offers hardware infrastructure like switches, routers, servers, load balancers, firewalls, storage. Usually these resources are virtual resources if these are bare metal versions, then the term used is Metal as a Service (MaaS). Notable examples of IaaS are Amazon Web Services, Google Compute Engine, IBM Softlayer (which has both IaaS and MaaS offerings).

PaaS goes beyond IaaS and offers a computing platform, which includes managed operating system, execution environment, storage, database and HTTP server. The platform is managed by the provider, this allows application developers to build applications without the complexity and inherent cost of managing the underlying stack.

SaaS compared to PaaS or IaaS offers an application that is completely managed by the provider. Typical applications include databases, CRM software, Git repositories, etc. The pricing is usually pay-per-use or subscription based.

Modern day Cloud computing has adapted to the service oriented architecture by the means of microservices. Microservices are services that are isolated from each other and communicate over a network in order to fulfill a goal. The main difference between microservices and SOA is that the latter focuses on reusability and integrating larger business applications, while the former focuses on replacing an application with a set of services that can be replaced, updated and scaled independently. Each microservice can be implemented in different programming languages, databases and software environment thus increasing the development speed. Also in contrast with SOA each microservice defines its required resources. Microservices also pose the advantage that you can scale specific bottlenecks in your application, by assigning a different number of instances to each microservice. Architecturally speaking, microservices should be designed with fault in mind. If a microservice instance fails while processing a task, that task gets assigned to another instance of the same microservice.

For high availability application the placement of these instances is constrained by locality restrictions, for example cloud service providers like Amazon often provide multiple Availability Zones (AZs), especially designed for high availability application such that they have a small chance of failing simultaneously. A microservice scheduler must be able to balance the number of instances of a service across a number of AZs depending on the availability restrictions. Other restrictions imposed on microservices might be: data locality, specific machine requirements like virtualization mode used, presence of a certain generation of GPU or processor generation.

Scheduling microservices is very similar to the online multi-capacity bin packing problem, for which multiple algorithms exist. This problem was studied for scheduling Virtual Machines (VMs) and is sometimes combined with offline phases of the algorithm [2] for increased performance. For example Song et al. [3] presents an online algorithm for scheduling VMs with using as few servers as possible reaching a competitive ratio of 3/2. Another algorithm named HarmonicMix [4] improves on the previous work, reaching a competitive ratio of 4/3, meaning that the number of bins necessary is only 4/3 bigger than offline scheduling algorithms with infinite migration.

These algorithms make the assumption that all machines are equal, but in order to optimize for the smallest price, we cant make such an assumption. The nature of microservices, fault tolerance and scalability, make them able to be run on a cluster of Amazon Spot instances. Amazon Spot instances are spare virtual machine capacity auctioned off in real time by Amazon to the highest bidder. Amazon Spot prices are usually 1/4–1/6 lower than their OnDemand counterparts, thus bringing huge cost savings. Each type of instance in each availability zone has a dynamic price set by supply and demand. Thus some instance types might become unavailable for periods of time. Qu et al. [5] has shown how you can balance availability with price while using Amazon Spot Instances to run web applications on them, by over provisioning resources depending on the availability constraints of the application.

This chapter presents the general features of cloud systems and services in Sect. 2, the main evaluation metrics in Sect. 3, then in Sect. 4 address the SLA issue for Cloud Systems. Section 5 presents the modeling of Cloud Systems using CloudSim and the extension for it for scheduling algorithms.

2 General Features of Cloud Systems and Services

Cloud solutions allow users to access via Internet various types of resources such as existing applications in the Cloud, frameworks that can be used for development of custom built applications, access to Virtual Machines for installing operating systems and also storage and sharing solutions.

The Cloud is now a significant choice for multiple types of users, common individuals, scientists or technical users so large datasets are generated, and have to be processed. The scheduling algorithms used in Clouds can be improved to fit the new patterns of jobs and big data sets using hybrid approaches that will consider independent tasks, tasks with dependencies, asymptotic scale requests or smaller rates of arriving jobs [7–9]. All this algorithms are designed considering the main features of Cloud systems and services, which are presented in Table 1.

A brief overview of Cloud systems considering dimensions like type, model, locality, stakeholders, comparison with other models, benefits and future is presented in Fig. 1.

The recent Cloud computing paradigm was designed in order to provide end users and businesses with various advantages such as: self-service provisioning, broad network access, resource pooling, elasticity, measured service, pay per use [1]. This approach is based on utility computing were we have infinite resources (as much as you need) and a concrete billing model (e.g. hourly).

The main benefits of Cloud systems are represent by the possibility to use high-scale/low-cost providers, by having any time/place access via a web browser, rapid scalability (incremental cost and load sharing), and a great focus on local IT systems.

We still have several concerns and open issues for Cloud systems, like performance evaluation, reliability and interoperability assurance, SLA negotiation, control of data

Table 1 General features of cloud systems and services

Feature	Description
Availability	Degree to which a system is in a specified state
Reliability	Power to remain functional with time without
Efficiency	The ratio of the useful work performed by a system to the total energy expended or heat taken in
Reusability	The level to which a component may be used in a number of systems or applications
Interoperability	The capability to integrate with different standards and technologies
Adaptability	The level of efficiency in adjusting a solution for the utilization in different context
Usability	The quantity to which a Cloud service could be used by particular consumers to gain certain aims with usefulness
Modifiability	The capability to make modifications to a component rapidly and cost effectively
Sustainability	Environmental effect of the Cloud service (usual carbon footprint or even energy capable of the Cloud services)
Scalability	The capability of a system to handle a growing amount of resources and workloads
Elasticity	"The degree to which a system is able to adapt to workload changes by provisioning and de-provisioning resources in an autonomic manner, such that at each point in time the available resources match the current demand as closely as possible" [6]

and offered service parameters, no standard API (a mixture of SOAP and REST and other standards), and many open issued about privacy, security, and trust.

We classify the main characteristics and issues about Cloud system considering nonfunctional aspects, economic models and technological features. This synthetic approach is presented in Fig. 2.

3 Evaluation Metrics

The evaluation metrics are presented for all described features in the previous section. A comprehensive and well described taxonomy of evaluation metrics where presented in [10]. According with this evaluation we have *basic performance metrics* (execution time, speedup, efficiency, scalability, elasticity, etc.), *Cloud capabilities* (latency, throughput, bandwidth, recoverability, storage capacity, software tunning,

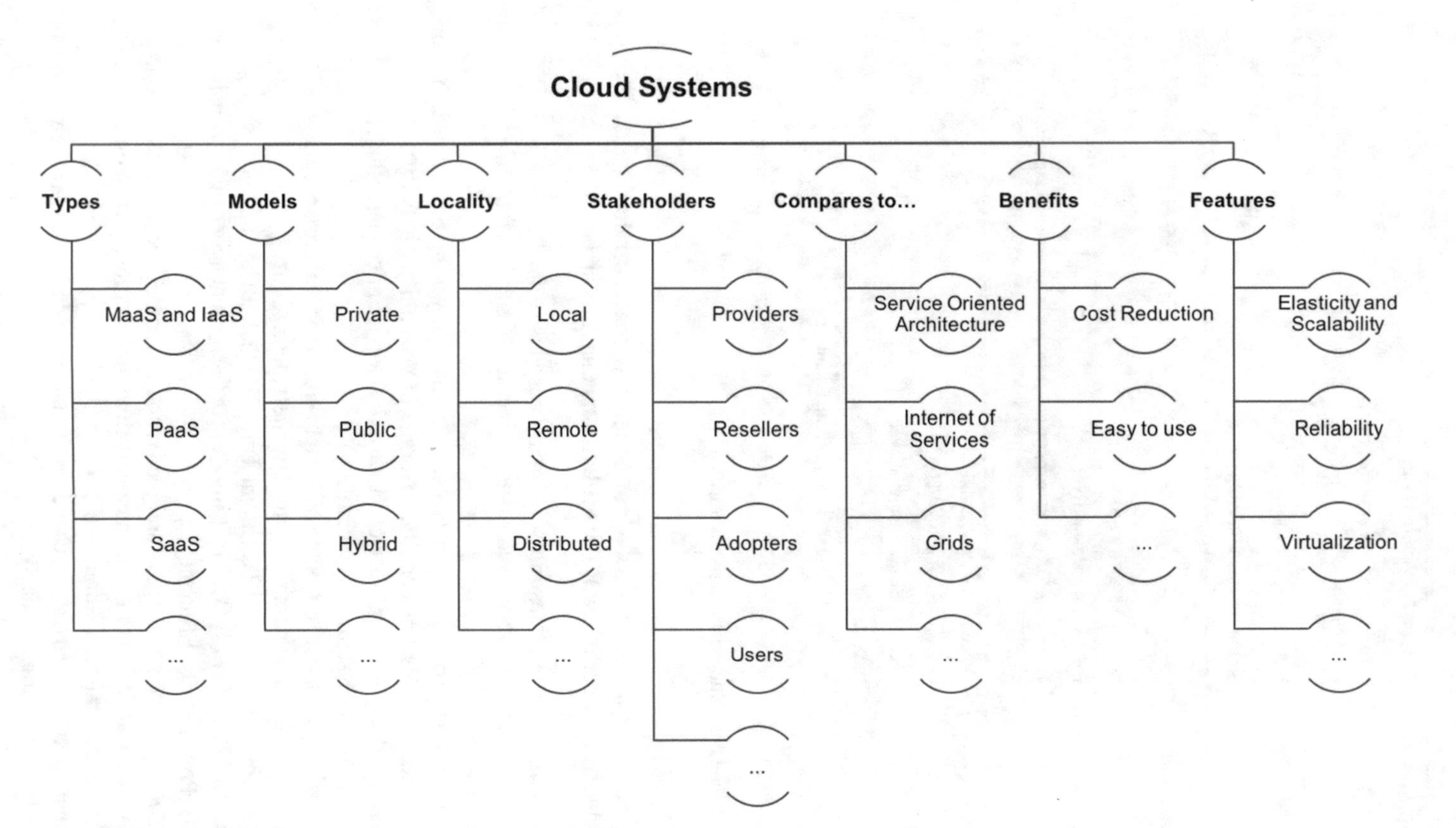

Fig. 1 Brief overview of Cloud Systems

Nonfunctional	Elasticity (ex: Amazon EC2)
	Reliability (ex: Vmware ecosystem)
	Quality of Service (ex: Amazon S3)
	Agility and adaptability (ex: FlexNet)
	Availability (ex: MS Azure)
Economic	Cost reduction
	Pay per use
	Improved time to market
	Return of investment (ROI)
	Turning CAPEX (capital expenditure) into OPEX (operational expenditure)
	Going Green
Technological	Virtualization (ex: Virtual Box)
	Multi-tenancy (ex: MS SQL)
	Security, privacy and compliance
	Data Management (ex: WebSphere)
	APIs and / or Programming Enhancements (ex: Hadoop)
	Tools

Fig. 2 Cloud computing characteristics/issues

etc.), and *Cloud productivity* (QoS, power demand, cost of services, availability, productivity, SLA, security, etc.). The **evaluations metrics** can be grouped by [11–13]:

- *Availability metrics*: "flexibility, accuracy, response time";
- *Reliability metrics*: "service constancy, accuracy of service, fault tolerance, maturity, recoverability";
- *Efficiency metrics*: "utilization of resource, ratio of waiting time, time behavior";
- *Reusability metrics*: "readability, coverage of variability, publicity";
- *Interoperability metrics*: "service Modularity, service interoperability, LISI (Level of Information System Interoperability)";
- *Adaptability metrics*: "coverage of Variability, other performance metrics";
- *Usability metrics*: "operability, attractiveness, learnability";
- *Modifiability metric*: "MTTC (Mean Time To Change)";
- *Sustainability metrics*: "DPPE (Data Centre Performance per Energy) parameter, PUE (Power Usage Efficiency)";
- *Scalability metric*: "average of assigned resources among the requested resources";
- *Elasticity metrics*: "boot time, suspend time, delete time, provision (or Deployment) time, total acquisition time";
- *Communication metrics*: "packet loss frequency, connection error rate, transfer bit/Byte speed, transfer delay";

- *Computation metrics*: "CPU Load, benchmark OP (FLOP) rate, instance efficiency (% CPU peak)";
- *Storage metrics*: "response time, latency, bandwidth, capacity";
- *Memory metrics*: "mean hit time, memory bit/Byte Speed, random memory update rate, response time (ms)";
- *Time metrics*: "computation time, communication time";
- *Data Security metrics*: "Is SSL applicable, communication latency over SSL, auditability, resistance to attacks";
- *Authentication metrics*: "meaning, sensitivity, effectiveness, confidentiality".

Other evaluation metrics can be defined to evaluate task scheduling and resource allocation systems [14–16]. We highlights here several performance evaluation metrics for a set of N jobs that is subject to a scheduling algorithm or policy in a Cloud system:

$$AverageWaitTime = \frac{1}{N} \sum_{j \in Jobs} (StartTime_j - SubmitTime_j).$$

$$AverageTurnaroundTime = \frac{1}{N} \sum_{j \in Jobs} (EndTime_j - SubmitTime_j).$$

$$FractionOfJobsTransferred = \frac{NumberOfJobsMigrated}{TotalNumberOfJobs}.$$

$$FractionDataVolumeTransferred = \frac{\sum_K (InputSize_K + OutputSize_K)}{\sum_J (InputSize_J + OutputSize_J)}.$$

$$DataMigrationOverhead = \frac{TotalDataMigrationTime}{\sum_J (EndTime_J - QueueTime_J)}.$$

These are several composed metrics defined for a task scheduling systems, but we can define many other evaluation metrics, according with the defined model and properties.

4 Performance and Service Level Agreement

One of the most important constraints of resource allocation techniques in the Cloud is the level of client satisfaction. This level is described by the Service Level Agreement (SLA) contract, which represents as a service level warranty between the provider and the customer of the service. Usually, some of the most important goals of the SLA contract are given by the necessity to have a common language between the customer and the services provider, and to verify the level of customer satisfaction during the use of the agreed services. An SLA contract is designed and planned based on the objective requests related to the cost reduction, efficiency increase and high performance, availability and highest level of security of the provider Cloudbased services.

Designing and implementing a SLA contract is a usual open discussion, because it often involves complex simulations or difficult results to analyze and implement. Our article has the purpose of presenting a survey of how Service Level Agreements (SLAs) are specified in Cloud computing environments. One of the methods for optimizing the resource allocation techniques is by satisfying the specifications of the SLA. The analysis of the level of satisfaction of the Service Level Agreement (SLA) and the improvement of the Quality of Service (QoS) is very important when studying those methods for optimizing the Cloud resource allocation.

When designing a Service Level Agreement contract in general we can discuss about the following phases [17–21] (see Fig. 3):

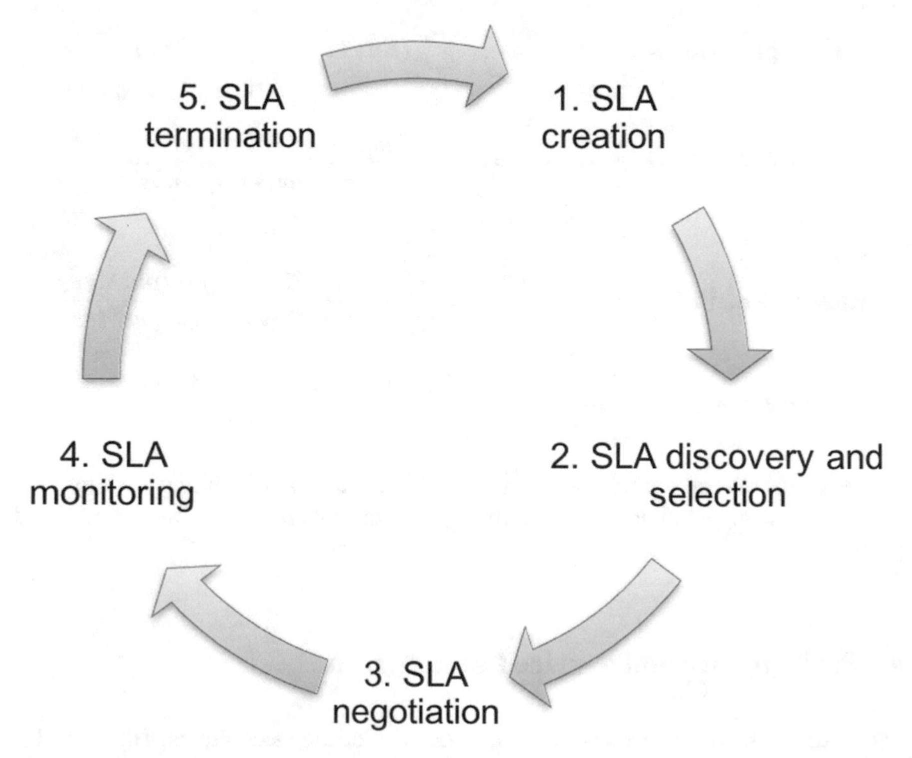

Fig. 3 SLA contract management phases

- The first phase in the design of an SLA contract is the SLA creation. During this phase service providers propose a SLA contract based on their capabilities and the contract contains SLA Offers. When discussing about the service consumers we refer to the SLA requirements specifications.
- The second phase in the design of an SLA contract is the SLA discovery and selection. During this phase, there exists a discovery of the offered services from different service providers and the selection of the services that satisfy both functional and non-functional requirements.
- The third phase in the design of an SLA contract is the SLA negotiation and it represents the negotiation and renegotiation step between providers and consumers.
- The fourth phase in the design of an SLA contract is the SLA monitoring phase when the service is starting and is provided to the consumer. During this phase the consumer monitors and validates the service characteristics offered by the service provider.
- The fifth phase when discussing about an SLA contract is the SLA termination which occurs when the SLA contract expires or either the consumer or the provider decide to end the agreement.

Cloud computing systems (or hosting datacenters) represent one of the main research areas in the field of distributed systems. Utility computing, reliable data storage, and infrastructure-independent computing are example applications of such systems [22].

Because adopting the Cloud services has various reasons such as lower costs because of the economy of resources (in comparison with when the client has to buy the necessary resources for example performant servers, etc.), transferring the responsibility when discussing the availability, maintenance, backup and lower licensing prices of the applications that are in place in the Cloud [23]. On the other hand, the fact that when adopting the Cloud computing paradigm with different purposes some of the service characteristics is transferred out of the customer control to the Cloud computing services providers. For this reason, there is a need for a contract (Service Level Agreement (SLA)) between the customer and the provider [24].

In addition, one of the characteristics of the contract is represented by the profit in the system, which depends on how the system can meet the SLA. (e.g. average response time, number of jobs completed per unit of time, availability of the resources in the system etc.). The SLA contract usually specifies the constraints that need to be satisfied by the system in order to achieve the level of client satisfaction and system quality of service (QoS) agreed in this contract. Another way of thinking about a Service Level Agreement (SLA) contract is that it represents a complex document that describes the parameters that need to be satisfied in the time and at the values described by the range specified in the Service Level Agreement.

In the following section of this article we present int three Tables the Cloud SLAs which are drafted by Cloud providers. In this section we introduce Service Level Agreement (SLA) and Service level ranking criterion (Table 2) based on the JDN/CloudScreener/Cedexis U.S. Cloud Benchmark March 2016.

Table 2 Service level agreement and service level ranking criterion

	AWS	Google	Microsoft	Rackspace	IBM softlayer
Announced SLA	99.95%	99.95%	99.95%	99.90%	99.73%
Service level ranking	90.83	90.83	79,38	73,33	73,21

The service level index is based on different qualitative criterion such as the geographical coverage (presence in the U.S. and outside of the U.S.), the number of certifications, the SLA, and the range of VM (the full ranking methodology is available on the JDN/CloudScreener/Cedexis U.S. Cloud Benchmark website: http://www.journaldunet.com/us-Cloud-benchmark/) [25]. This article presented a survey of the existing major Cloud providers and how Service Level Agreements (SLAs) have been defined by these providers.

SLAs are very important in utility computing systems because they characterize the various interactions between the Cloud providers and the clients or consumers. The future research in SLA-oriented Cloud computing has to take into account the following goals:

- the service management has to be based on the requested levels of service characteristics; the characteristics that need to be taken into consideration when designing a Service Level Agreement (SLA) are related both to the computational risks and the service requirements;
- to identify the execution risks involved in the execution of applications, risks that might have an impact on the levels of performance specified in the Service Level Agreements (SLAs);
- there has to exist an equilibrium between the customer satisfaction and the level of provider profit;
- there might be a need to model the different resource management designs that are based both on the customers service demands and existing service properties;
- there are various operations that need to be taken into consideration when decinding to construct a Service Level Agreements contract such as: discover-service provider, define-SLA elements, establish-agreement, monitor SLA violation, terminate-SLA and SLA violation control [17].

Currently, the automatic negotiation of a Service Level Agreement in Cloud computing is still an open issue. Other open issues are scalability and heterogeneity of a service in Cloud computing, dynamic environmental changes, multiply QoS parameters and SLA suitable for cross domains [17]. Finally, we need to take into consideration the fact that the SLA needed in order to define the trust and quality of services has to be based on an agreed framework that represents a contract between consumer and provider about service terms such as: performance, availability and billing [26]. All these challenges are still open and can be explored in the future.

5 Modeling of Cloud Systems Using CloudSim

The CloudSim [27] Java toolkit allows the modeling of different entities in a Cloud environment and simulate various scenarios: evaluate the configuration of a Cloud System, resource allocation policies or scheduling algorithms. CloudSim is an extensible framework (developed in CLOUDS Laboratory, Computer Science and Software Engineering Department of the University of Melbourne) due to its high modularity. The Cloud entities: data centers, hosts, VMs, jobs, inter-host agreements or VM allocation policies are modeled as classes in different packages that can be interconnected or extended to enhance them with additional functionality. A quick look over running CloudSim environments in Eclipse is presented in Fig. 4. CloudSim has the main benefit by having cloud resource provisioning modules, energy-efficient management of data center resources strategies and support for optimization.

In this section we analyze the required steps to extend the CloudSim framework for implementing a custom scheduling algorithm and evaluate the simulation results and algorithm performance.

Cloudlet represents the abstraction of a Job/Task. Some of its properties are the length (computational requirements) and file size (IO), either for input or output. For statistics computation, it stores the VM that executed it, the start and finish execution times.

We will extend the Cloudlet by creating an additional object: Task that stores application specific information, and is connected to a Cloudlet object using the same id value.

```
1  public class Task {
     // rank the current task reported to the complete set of tasks
3    public double processingRank;
     // rank the current task reported to the complete set of tasks
5    public double ioRank;
     // connect the Task to a Cloudlet using this attribute
7    public int id ;
     public long length;
9    public long fileSize;
     public long outputSize;
11   public int pesNumber;
     public long deadline;
13   public long io;
```

Extend a Cloudlet using the Task class.

A *VM* object holds the properties of the underlying hardware, a link to the physical host and the policy for submitting tasks on PEs. We can extend the VM with the Resource object and add the specific attributes required by the scheduling algorithms, in our case, we added the resource load attribute.

```
1  public class Resource {
      // connect the Resource to a VM using this attribute
3    private int id;
     // work load of the current VM
```

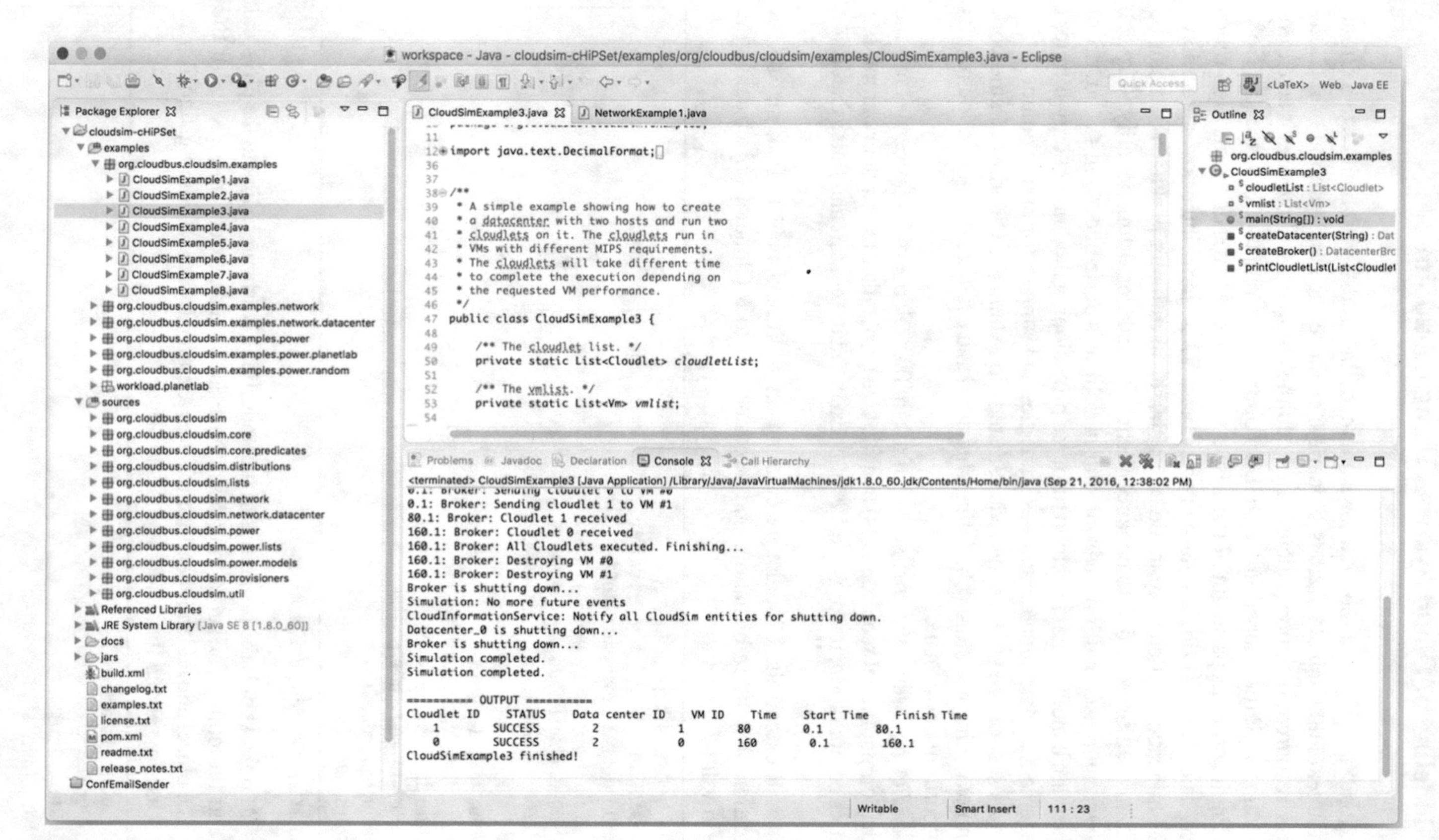

Fig. 4 Quick look over running CloudSim environment in Eclipse

```
5   private int load;
    public ArrayList<Task> schedTasks = new ArrayList<Task>();
7   private int mips;
    private int ram;
9
    public Resource(int id, int load, int mips, int ram) {
```

Extend a VM using the Resource class.

The *DatacenterBroker* handles the allocation of Cloudlets on VMs using the function *submitCloudlets*. The implementation of a scheduling algorithm can be done by extending this class and overwriting the submitCloudlets function.

```
1   protected void submitCloudlets() {
      int vmIndex = 0;
3     for (Cloudlet cloudlet : getCloudletList()) {
       Vm vm;
5       // if user didn't bind this cloudlet and it has not been executed yet
        if (cloudlet.getVmId() == -1) {
7         vm = getVmsCreatedList().get(vmIndex);
        } else { // submit to the specific vm
9         vm = VmList.getById(getVmsCreatedList(), cloudlet.getVmId());
          if (vm == null) { // vm was not created
11          Log.printLine(CloudSim.clock() + ": " + getName() + ": Postponing
                 execution of cloudlet "
               + cloudlet.getCloudletId() + ": bount VM not available");
13          continue;
          }
15      }

17      Log.printLine(CloudSim.clock() + ": " + getName() + ": Sending cloudlet "
            + cloudlet.getCloudletId() + " to VM #" + vm.getId());
19      cloudlet.setVmId(vm.getId());
        sendNow(getVmsToDatacentersMap().get(vm.getId()),
21          CloudSimTags.CLOUDLET_SUBMIT, cloudlet);
        cloudletsSubmitted++;
23      vmIndex = (vmIndex + 1) % getVmsCreatedList().size();
        getCloudletSubmittedList().add(cloudlet);
25    }

27    // remove submitted cloudlets from waiting list
      for (Cloudlet cloudlet : getCloudletSubmittedList()) {
29      getCloudletList().remove(cloudlet);
      }
31  }
```

Default scheduling in DatacenterBroker.

```
1 public class Scheduler  extends DatacenterBroker implements IScheduler{

3   public static SchedulingMethods method;
    public static ArrayList<Task> tasks;
5   public static ArrayList<Resource> resources;
```

```
7   @Override
    protected void submitCloudlets() {
9     switch(method){
      case Default: defaultSchedule();
11      break;
      case SJF: sjf();
13      break;
      case ClusteringSJF: clusteringSJF();
15      break;
      }
17  }
```

Extend a DatacenterBroker using the Scheduler class.

```
    Collections.sort(tasks, new Comparator<Task>() {
2
      @Override
4     public int compare(Task o1, Task o2) {
        int r = (int)(o1.length - o2.length);
6       return r != 0 ? r : (int)(o1.io - o2.io);
      }
8   });

10  for (int i = 0; i < tasks.size(); i++) {
      int id = tasks.get(i).id;
12    Cloudlet cloudlet = cloudletList.get(id);
      Vm vm;
14    // if user didn't bind this cloudlet and it has not been executed yet
      if (cloudlet.getVmId() == -1) {
16      vm = getVmsCreatedList().get(vmIndex);
      } else { // submit to the specific vm
18      vm = VmList.getById(getVmsCreatedList(), cloudlet.getVmId());
        if (vm == null) { // vm was not created
20        Log.printLine(CloudSim.clock() + ": " + getName() + ": Postponing
              execution of cloudlet "
            + cloudlet.getCloudletId() + ": bount VM not available");
22        continue;
        }
24    }

26    cloudlet.setVmId(vm.getId());
      sendNow(getVmsToDatacentersMap().get(vm.getId()), CloudSimTags.
          CLOUDLET_SUBMIT, cloudlet);
28    cloudletsSubmitted++;
      vmIndex = (vmIndex + 1) % getVmsCreatedList().size();
30    getCloudletSubmittedList().add(cloudlet);
    }
32
    // remove submitted cloudlets from waiting list
34  for (Cloudlet cloudlet : getCloudletSubmittedList()) {
```

Implement the SJF algorithm.

6 Conclusion

The new trends in modeling and simulation of Cloud Systems require performance evaluation metrics with a high level of accuracy. We presented in this chapter several feature of Cloud systems and services ans a set of evaluation metrics. We included a practical example using CloudSim. which analyze the required steps to extend the CloudSim framework for implementing a custom scheduling algorithm and evaluate the simulation results.

Acknowledgements The research presented in this paper is supported by the projects: *DataWay*: Real-time Data Processing Platform for Smart Cities: Making sense of Big Data—PN-II-RU-TE-2014-4-2731; *MobiWay*: Mobility Beyond Individualism: an Integrated Platform for Intelligent Transportation Systems of Tomorrow - PN-II-PT-PCCA-2013-4-0321; and *cHiPSet*: High-Performance Modelling and Simulation for Big Data Applications, ICT COST Action IC1406.

We would like to thank the reviewers for their time and expertise, constructive comments and valuable insight.

References

1. Mell, P., Grance, T.: The NIST definition of cloud computing. Commun. ACM **53**(6), 50 (2010)
2. Leinberger, W., Karypis, G., Kumar, V.: Multi-capacity bin packing algorithms with applications to job scheduling under multiple constraints. In: Proceedings of 1999 International Conference on Parallel Processing, pp. 404–412. IEEE (1999)
3. Song, W., Xiao, Z., Chen, Q., Luo, H.: Adaptive resource provisioning for the cloud using online bin packing. IEEE Trans. Comput. **63**(11), 2647–2660 (2014)
4. Kamali, S.: Efficient bin packing algorithms for resource provisioning in the cloud. In: Algorithmic Aspects of Cloud Computing, pp. 84–98. Springer (2016)
5. Qu, C., Calheiros, R.N., Buyya, R.: A reliable and cost-efficient auto-scaling system for web applications using heterogeneous spot instances. J. Netw. Comput. Appl. **65**, 167–180 (2016)
6. Herbst, N.R., Kounev, S., Reussner, R.: Elasticity in cloud computing: what it is, and what it is not. In: Proceedings of the 10th International Conference on Autonomic Computing (ICAC 13), pp. 23–27 (2013)
7. Vasile, M.A., Pop, F., Tutueanu, R.I., Cristea, V., Kołodziej, J.: Resource-aware hybrid scheduling algorithm in heterogeneous distributed computing. Future Gener. Comput. Syst. **51**, 61–71 (2015)
8. Vasile, M.A., Pop, F., Tutueanu, R.I., Cristea, V.: HySARC2: hybrid scheduling algorithm based on resource clustering in cloud environments. In: International Conference on Algorithms and Architectures for Parallel Processing, pp. 416–425. Springer (2013)
9. Sfrent, A., Pop, F.: Asymptotic scheduling for many task computing in big data platforms. Inf. Sci. **319**, 71–91 (2015)
10. Hwang, K., Bai, X., Shi, Y., Li, M., Chen, W.G., Wu, Y.: Cloud performance modeling with benchmark evaluation of elastic scaling strategies. IEEE Trans. Parallel Distrib. Syst. **27**(1), 130–143 (2016)
11. Bardsiri, A.K., Hashemi, S.M.: Qos metrics for cloud computing services evaluation. Int. J. Intell. Syst. Appl. **6**(12), 27 (2014)
12. Kan, S.H.: Metrics and Models in Software Quality Engineering. Addison-Wesley Longman Publishing Co., Inc. (2002)
13. Iosup, A., Ostermann, S., Yigitbasi, M.N., Prodan, R., Fahringer, T., Epema, D.: Performance analysis of cloud computing services for many-tasks scientific computing. IEEE Trans. Parallel Distrib. Syst. **22**(6), 931–945 (2011)

14. Topcuoglu, H., Hariri, S., Wu, M.Y.: Performance-effective and low-complexity task scheduling for heterogeneous computing. IEEE Trans. Parallel Distrib. Syst. **13**(3), 260–274 (2002)
15. Feitelson, D.G., Rudolph, L.: Metrics and benchmarking for parallel job scheduling. In: Workshop on Job Scheduling Strategies for Parallel Processing, pp. 1–24. Springer (1998)
16. Pop, F., Cristea, V., Bessis, N., Sotiriadis, S.: Reputation guided genetic scheduling algorithm for independent tasks in inter-clouds environments. In: 27th International Conference on Advanced Information Networking and Applications Workshops (WAINA), pp. 772–776. IEEE (2013)
17. Wu, L., Buyya, R.: Service Level Agreement (SLA) in Utility Computing Systems. IGI Global (2012)
18. Debusmann, M., Keller, A.: SLA-driven management of distributed systems using the common information model. In: Integrated Network Management VIII, pp. 563–576. Springer (2003)
19. Alhamad, M., Dillon, T., Chang, E.: SLA-based trust model for cloud computing. In: 13th International Conference on Network-Based Information Systems (NBIS), pp. 321–324. IEEE (2010)
20. Venticinque, S., Aversa, R., Di Martino, B., Rak, M., Petcu, D.: A cloud agency for SLA negotiation and management. In: European Conference on Parallel Processing, pp. 587–594. Springer (2010)
21. Sahai, A., Machiraju, V., Sayal, M., Van Moorsel, A., Casati, F.: Automated SLA monitoring for web services. In: International Workshop on Distributed Systems: Operations and Management, pp. 28–41. Springer (2002)
22. Goudarzi, H., Ghasemazar, M., Pedram, M.: SLA-based optimization of power and migration cost in cloud computing. In: 2012 12th IEEE/ACM International Symposium on Cluster, Cloud and Grid Computing (CCGRID 2012), pp. 172–179. May 2012
23. Goudarzi, H., Pedram, M.: Multi-dimensional SLA-based resource allocation for multi-tier cloud computing systems. In: 2011 IEEE International Conference on Cloud Computing (CLOUD), pp. 324–331. IEEE (2011)
24. Dastjerdi, A.V., Tabatabaei, S.G.H., Buyya, R.: A dependency-aware ontology-based approach for deploying service level agreement monitoring services in cloud. Softw Pract. Experience **42**(4), 501–518 (2012)
25. CCMBenchmark: JDN, CloudScreener, Cedexis US. Cloud benchmark website. http://www.journaldunet.com/us-cloud-benchmark (2016)
26. Alhamad, M., Dillon, T., Chang, E.: Conceptual SLA framework for cloud computing. In: 4th IEEE International Conference on Digital Ecosystems and Technologies, pp. 606–610. IEEE (2010)
27. Calheiros, R.N., Ranjan, R., Beloglazov, A., De Rose, C.A., Buyya, R.: CloudSim: a toolkit for modeling and simulation of cloud computing environments and evaluation of resource provisioning algorithms. Softw Pract Experience **41**(1), 23–50 (2011)

Science Gateways in HPC: Usability Meets Efficiency and Effectiveness

Sandra Gesing

1 Introduction

HPC (High-Performance Computing) infrastructures provide the means for compute-intensive modeling and simulations to achieve results in reasonable time. Efficiency and effectiveness are the traditional key targets for the optimization of such applied scientific methods and they are major drivers for research and developments in HPC. In the last years a further target has arisen driven by the needs of user communities to enable them to focus on their research questions without becoming deeply acquainted with the complex technical details of HPC: usability of modeling and simulations in HPC. Science gateways address this aspect as end-to-end solutions providing intuitive user interfaces while connecting to the underlying complex infrastructures and hiding the technical details as far as feasible and desired from the users. This trend is reflected in quite a few web development frameworks, containerizations, science gateway frameworks and APIs with different foci and strengths, which have evolved to support the developers of science gateways in implementing an intuitive solution for a target research domain. Science gateways have evolved into a new era since 2014 when providers of distributed computing infrastructures reported the first time that the computing and storage resources have been applied more often via science gateways than via command line [1]. Part of this success can be credited to the development of reliable and mature science gateway frameworks over the last decade [2]. Especially the rise of larger data amounts and the importance of workflows for user communities have been recognized and sophisticated data and workflow management solutions [3] have found their way into such frameworks.

The challenges for developers of specific science gateways, which apply HPC infrastructures, are manifold: from intuitive user interfaces for the targeted research domain and security features through efficient job, data and workflow management to

S. Gesing (✉)
University of Notre Dame, Notre Dame, USA
e-mail: sandra.gesing@nd.edu

J. Kołodziej et al. (eds.), *Modeling and Simulation in HPC and Cloud Systems*,
Studies in Big Data 36, https://doi.org/10.1007/978-3-319-73767-6_5

parallelization of applications employing parallel and distributed architectures. The knowledge about existing science gateway technologies and their distinctive features and strengths helps developers to select a suitable framework or API without the need to re-invent the wheel and to start the development of a specific solution from scratch.

In the area of science gateways several sources are available to get a well-informed impression of the state-of-the-art technologies and novel developments. Yearly science gateway workshops are established in Europe, the US and Australia, which are partnering and form an international platform to shape future directions for research, foster the exchange of ideas, standards and common requirements and push towards the wider adoption of science gateways in science [4, 5]. The peer-reviewed publications of the workshops and the joint special issues reflect the international standard in this field [6]. IEEE has also observed the importance of science gateways and the IEEE Technical Area on Science Gateways is a further source of information on events, publications and projects [7]. Besides such community-driven resources, the US National Science Foundation (NSF) [8] - as one of the main funding bodies in the US - has recognized the significance of science gateways and is funding the Science Gateways Community Institute [9]. The Science Gateway Institute provides among other services an excellent contact for general information on projects and technologies [5]. The selection of a suitable technology for a specific use case is essential and helps reducing the effort in implementing a science gateway by reusing existing software or frameworks. Thus, a solution for a user community can be provided more efficiently. Additionally, novel developments in web-based technologies and agile web frameworks allow for supporting developers in efficiently creating web-based science gateways.

2 Science Gateways and Usability

The overall goal of science gateways is to provide an end-to-end solution and increase the usability of applications especially for researchers who are not necessarily IT specialists. The significance of usability and graphical user interfaces is evident in the history of IT developments in the last 50 years: Doug Engelbart's Augmentation of Human Intellect project, which developed a mouse-driven cursor and multiple windows in the 60s [10], Apple's designs starting in the 70s and resulting in a hype in the last 10 years around smartphones and tablets, the first web browser [11] and an ISO standard on usability for "visual display terminals" in the 90s [12]. The Internet revolutionized research in the last 25 years with increasingly more sophisticated and efficient distributed computing infrastructures and data management solutions having evolved to maintain and increase their usability. Novel developments in web-based technologies as well as agile web frameworks allow for supporting developers in efficiently creating user interfaces for web-based science gateways.

On the user interface side many libraries and frameworks have evolved and we will only mention a few without the claim of completeness. In general, JavaScript libraries, CSS and HTML5 with Ajax [13] allow for dynamic websites focusing on the frontend with advanced features. jQuery [14] is a widely used JavaScript library with standard user interface methods for HTML document traversal and manipulation and event handling. jsPlumb [15] is also a JavaScript library with focus on the illustration of graphs and workflows with many implemented features for the appearance of nodes and edges and corresponding annotations. 3D graphics can be seamlessly created and edited in web browsers via the JavaScript API WebGL [16] without the need of installation of further software. The front-end framework Semantic UI [17] makes use of JavaScript library jQuery, while providing intuitive classes for designing web user interfaces based on the philosophy "everything arbitrary is mutable". The web application frameworks ReactJS [18], Foundation [19], AngularJS [20] uses declarative programming and follows the MVC concept (Model-View-Controller) [21] to separate data, presentation and logical components in a clean design.

While the look-and-feel of the user interface is especially important for the acceptance in the user community, the backend and the integration with the underlying infrastructure, which is mostly hidden from the users, is the more complex task from the technical point of view. Some technologies are widely used for web-based science gateways but are lacking standard libraries for the support of HPC infrastructures such as the open source content management systems Drupal [22] and Joomla [23] and the high-level framework Django [24]. Thus, the developers are creating such integrations from scratch. The lack of HPC support out of the box also applies to portal frameworks such as Liferay [25] and Pluto [26] but offering the advantage of re-usability of so-called portlets. The portal frameworks are implementations of the JSR168/JSR286 [27, 28] standards and they enable to implement portlets once and deploy them in every portal framework, which supports these standards. Especially Liferay is widely used for science gateways in the HPC community. In the last eight years a couple of science gateway frameworks have been developed on top of Liferay, benefitting from the available authentication and authorization mechanisms and layout features, e.g., gUSE/WS-PGRADE [29].

In general, the architecture of science gateway technologies for distributed systems consists of four layers: (1) the user interface layer, (2) the application layer, (3) the high-level services layer such as job management and data management and if applicable workflow management and (4) the connected cluster, grid and cloud infrastructures (see Fig. 1). While the first two layers may be specific for each science gateway developed via a science gateway technology, the third and fourth layer are generic and can be re-used for any science gateway irrespective of its target domain. The generic requirements on such layers have led to the development of multiple mature science gateway technologies. We refer to examples here, which are free and available as open source and can be as such further developed by the community and are not based exclusively on a business model.

One category includes workbenches such as Taverna [30], the Kepler workbench [31], KNIME (the Konstanz Information Miner) [32], and the UNICORE Rich Client [33]. These examples are additionally offering workflow management capabilities. They necessitate the installation of software on the user side and offer a workflow canvas to graphically create and edit workflows and submit them to the underlying infrastructure. Each user interface layer provides the same look-and-feel for all

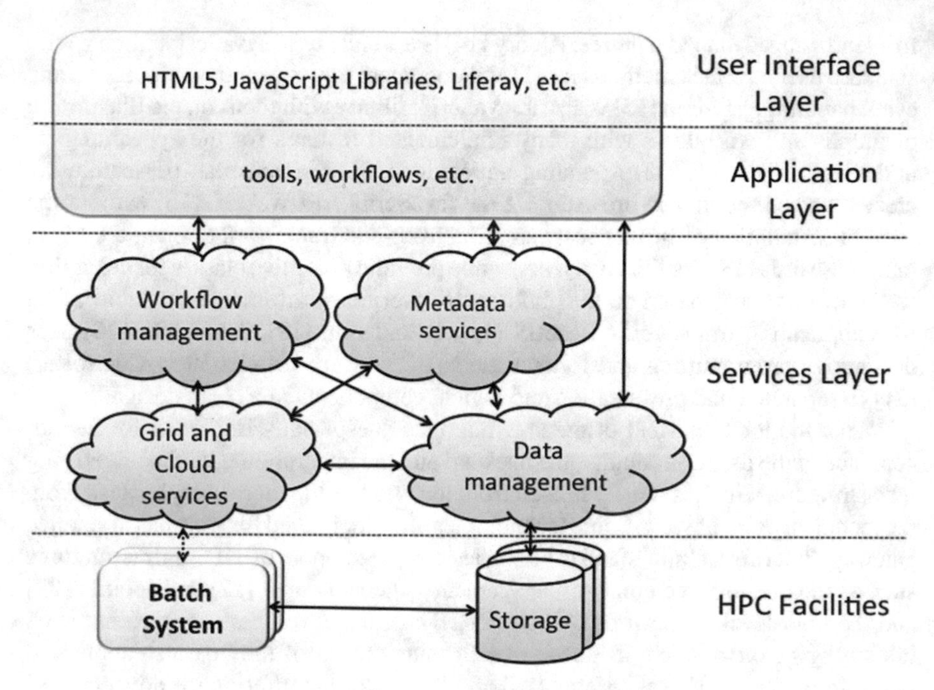

Fig. 1 The general infrastructure for science gateways with providing access to HPC resources. For each layer are examples of technologies provided

applications. The target infrastructures are quite different for these examples though. Taverna supports the workflow management of available web services, whereas Kepler targets command line tools like R scripts or compiled C implementations. KNIME is intuitive, versatile, widely used, and currently being extended to support generic access to HPC resources. The UNICORE Rich Client focuses on the exploitation of compute and data infrastructures, which are integrated via the UNICORE grid middleware [34].

The second category of science gateways contains web-based science gateway frameworks. gUSE/WS-PGRADE, Galaxy [35], HUBzero [36] and the Catania Science Gateway Framework [37] belong to this category. The first two offer workflow editing features and workflow management as well whereas HUBzero provides workflow management options in the backend via the workflow management system Pegasus [38] but focuses more on the integration of single applications and collaboration tools analogous to the Catania Science Gateway Framework. gUSE/WS-PGRADE and Galaxy offer generic workflow canvasses capable of managing command line tools and web services. The concepts behind creating the workflows are quite different though. While WS-PGRADE includes the option for the users to upload and invoke scripts and computational tools, Galaxy is designed as toolbox, which is configured by an administrator and users can select from a list of available tools. The extension of the science gateway frameworks with user interfaces especially tailored to a specific application can be performed in WS-PGRADE as portlets developed on top of Liferay. Galaxy is not developed on top of a standard framework and thus does not directly support the implementation of specific user interfaces but since it is available as open source, developers are able to extend the framework to communities' demands.

The third category is concerned with the development of science gateways and includes mature APIs and libraries offering features for the implementation of the first three layers of the science gateway architecture. Examples are Apache Airavata [39], the Agave Platform [40] and the Vine Toolkit [41], which aim at reducing the effort on the developer side while enabling to apply novel user interface technologies and frameworks. All three frameworks are supporting diverse programming languages and the basic concept is the same.

3 Designing Science Gateways

The close collaboration with the respective user community is crucial to gather all necessary information and requirements on a science gateway that is intended to serve for the specific use case. This usually underestimated design task is more often than not the most challenging one. While users are mainly experts in their research domain, they may be not aware of the implications of using specific software, the availability of a computational tool, security demands or concepts such as workflows. The exact layout for the science gateway is usually a continuous and iterative process with suggestions from developers for the layout and feedback and comments from the user community.

Through the experience with several projects and communities, we have created a checklist for important topics to discuss and address in collaboration with the communities [42]. This checklist can be used for supporting the creation of a Software Requirement Specification (SRS), for example, following the 830-1998 - IEEE Recommended Practice for Software Requirements Specifications [43]. According to the recommendation a SRS should be

(a) Correct;
(b) Unambiguous;
(c) Complete;
(d) Consistent;
(e) Ranked for importance and/or stability;
(f) Verifiable;
(g) Modifiable;
(h) Traceable.

It should address the software product—not the process of producing a software product. In the MoSGrid project (Molecular Simulation Grid) [44], for example, we created first a survey answered by about 50 domain researchers in computational chemistry, with questions such as which tools they use, whether they know about computational workflows and may use already workflows and whether they would like to share their data and/or workflows. The domain partners of the project (five domain researchers from three affiliations) have been directly involved in the design process.

The topics can be distinguished in the three main categories: domain-specific topics, organizational topics and technical topics. Domain-specific topics can be again divided in three major groups:

1. Requirements, which lay in the nature of the research topic.
2. Requirements, which refer directly to the target community and their specific needs, their diversity in experience and knowledge regarding the research topic and/or computational tools and their analysis steps.
3. Requirements, which result from available specific resources from lab instruments to local, on-campus, national to international distributed computing infrastructures.

Topics of the three groups are in detail explained in Table 1.

In contrast to domain-specific topics, organizational topics refer to measures for a successful collaboration in general, which are influenced via external factors of a project such as time constraints of a grant or internal factors such as availability of alpha- or beta-testers (see Table 2). It may be not feasible to receive information on all topics or to set up all organizational aspects from the beginning but important is to raise the topics and start the conversation. The information is essential for the correct choice of technologies and may prevent a significant amount of refactoring. While in industry the creation of a system specification is the common approach to define all requirements, projects in academia work differently and are often more dynamic and less clearly specified. One reason lies in the research nature of the projects.

Besides the topics deriving from the research domain and the interdisciplinary collaboration, also technical topics of the design of available infrastructures and considerations regarding the involved development team or single developers are important to examine. Such subjective project analyses are complemented by the investigation of objective conditions such as available support of suitable technologies. See Table 3 for a comprehensive list of technical topics. This checklist is intended to support the design decision process and is general applicable. Each use case is different and there is not one technology, which fits them all but various mature solutions, which can form the basis for diverse science gateways.

4 Reusability of Scientific Methods and Reproducibility of Science

One of the goals of science gateways is to offer methodologies for performing analyses, which can be re-used for different data sets and by different users. Thus, many science gateways offer sharing possibilities within a community, between different

Table 1 The checklist illustrates the domain-specific groups and the topics, which can be used by principal investigators and/or developers for designing and implementation of a successful science gateway

Groups	Topics	Examples
Requirements referring to the research topic	Goal and target area of the envisioned science gateway	Workflows for computational drug design using docking tools
	Scale and format of the available data	Molecular structures in PDB format
Requirements referring to the target user community	Groups of users distinguished via their experience in the research domain	Novel users to the research topic such as students or experienced users in the research domain
	Groups of users distinguished via their experience with computational tools	Wet-lab researchers mostly familiar with working with Excel or researchers familiar with command line usage of computational tools
	Layout and feature requirements	Strictly pre-configured user interfaces, possibilities for changing parameter configurations or possibilities to process own scripts
	Priorities of features and options	A list ranging from must-have to great-to-have options
	Integration of existing applications or development of new applications from scratch	Computational tools already used in the community, e.g., Gromacs, or developing statistical approaches with R
	Visualization	Browsing of data or interactive modules like a molecule editor
	Workflow management	Pre-configuration of connected tasks for a certain purpose such as optimizing molecular structures for a docking workflow
	Security and privacy management	Private space for research results before publication or patents and shared spaces for results afterwards
Requirements referring to available infrastructures	Hardware	External disk at a lab containing the input data
	Credentials	Access via on-campus accounts or Grid certificates for national resources
	Batch systems, Grid middlewares or Cloud systems	PBS, UNICORE, etc.
	Data management systems	dCache, iRODS etc.

Table 2 The checklist illustrates the external and internal topics and measures on the organizational side, which can be used by principal investigators and/or developers for designing and implementation of a successful science gateway

Groups	Topics	Measures
External topics	Limited funding and time constraints	Project plan with deliverables and milestones
	Availability of data and computational tools from third party affiliations	Communicating via emails and calls with third party affiliations
Internal topics	Concurrent collection of requirements and features	Weekly meetings of researchers and developers or development team
	Concurrent feedback during the development	Agreement on alpha- and beta-testers in the community
	Location of teams	Distributed teams compared to local teams necessitate the use of conference calls as well as emails to a larger extent, maybe under consideration of different time zones

Table 3 The checklist illustrates the subjective and objective factors and the topics, which can be used by principal investigators and/or developers for designing a successful science gateway

Groups	Topics
Subjective factors	Experience with existing frameworks, programming languages and data access methods
	Effort for extending existing frameworks compared to novel developments for the specific use case
	Synergy effects with other science gateway projects
	Available infrastructure in the hosting environment including security infrastructure and resources
Objective factors	Available support of suitable technologies
	Scalability of suitable technologies
	Technologies of the applications, which have to be integrated
	Technical requirements of the applications and/or of access to input data
	Performance measures of applications

science gateway instances of one technology or even between diverse technologies, e.g., via the SHIWA platform [45, 46]. Even though such sharing options are available, reusability of methodologies and reproducibility of science are mainly dependent on two main aspects:

1. The willingness of researchers to share methodologies and data.
   Even with easy-to-use sharing options, researchers need to perform further steps to provide their methodologies and data to a community, which cost time and effort. They might have invested a large amount of time and computing power to create these and see an advantage to keep the knowledge in their group and between collaborators for creating further results with this part of their research. A survey in the MoSGrid [47] community elucidated that 70% would share their results and molecular structures in a repository after they have published them or own a patent. The disposition to share tools and workflows was higher with nearly 90%. If these results can be transferred to researchers in general, the 70% or 90%, respectively, would be a promising result to achieve reproducibility of science at a high rate.
2. Technical dependencies of methodologies and data.
   Methodologies are among others dependent on operating systems, tools in diverse versions and local or distributed data. A study on the social marketplace MyExperiment [48] for sharing Taverna workflows, for example, presents that only 20% of the workflows are reproducible and reusable out of the box. For solving such problems in science gateways, the different sharing possibilities have to be analyzed and – where necessary – tools, data and workflows have to be provided in diverse infrastructures and via various job and data management systems.

While research areas and the science gateway technologies are independent of institutional, state or national boundaries, this does often not apply to research infrastructures, which offer HPC infrastructures such as EGI [49] and PRACE [50] in Europe or XSEDE [51] in the US. The acquisition and maintenance of resources depend on funding, which can be institutional, national or on international level. Thus, the use of such resources is bound to policies and rules of the funding bodies. To support user communities across such boundaries, it is essential for science gateway creators to understand the effects of applying different research infrastructures.

Science gateways consists in general of three layers (see Fig. 1) and thus form a science gateway infrastructure:

1. User interface layer – This layer determines the layout and design of the user interface visible to the community.
2. Application layer – This layer is responsible for the features offered in the science gateway, e.g., generic applications such as security features for authentication to and authorization in the science gateway as well as domain-oriented applications such molecular structure editors. In case of workflow-enabled science gateways, this might be a workflow editor.

3. Services layer—This layer connects to services of the science gateway framework such as data repositories and services such as adaptors to apply batch, grid or cloud systems or distributed data management systems.

In a well-designed science gateway, the first layer is independent of the underlying research infrastructure while the latter influences the second and third layer.

Policies of research infrastructures add another complexity layer such as the application process to receive allocations on and access to available resources. The German National Grid Infrastructure, for example, only gives access to users of German universities and their collaborators. Other research infrastructures provide a more international approach such as InCommon [52], for example. InCommon is an international initiative for global interfederation for security credentials joined by over 40 national federations. However, policies of research infrastructures are quite diverse from each other, which thus hampers to determine generic challenges for policies of research infrastructures. The conclusion is that policies themselves can form a technical and organizational challenge dependent on restrictions resulting from them.

The following challenges in the application layer have to be considered.

A1. Security
Since one of the goals of science gateways is to create an easy-to-use interface to underlying resources, a single sign-on feature for accessing the science gateway via the same security credentials as the resources is highly desirable. Thus, the authentication mechanism of the science gateway or an additional feature in the science gateway has to support the credentials needed in a research infrastructure and corresponding authorization to compute and data resources.

A2. Available tools
Diverse research infrastructures have instantiated diverse policies regarding available tools. While some allow uploading own tools for submission to the research infrastructure, some only allow using pre-installed tools on the research infrastructure. In case of large software packages the installation has to be performed on the infrastructure for an efficient use. It is inefficient to install the whole package during each submission of a job referring to a tool in such a software package. Virtualizations like Docker containers [53] form efficient solutions here.

A3. Available data
Input data for a job or workflow may be available locally in the science gateway or in one research infrastructure. If the data is available in a certain research infrastructure and mandatory for the effective and efficient application of a tool or workflow in another research infrastructure, it needs to be added to the targeted infrastructure.

Similar challenges are faced in the services layer, though they take place on a different technical layer. Thus, the following challenges have to be addressed.

S1. Job management system
Each research infrastructure supports at least one job submission system. It might be a local, batch, grid or cloud system, which includes authentication, authorization and accounting mechanisms. The services available in the science gateway infrastructure have to be analyzed whether they support one of the available job management systems and its security demands. While diverse hardware architectures might be available in the research infrastructure, the differences are handled by the job management system.

S2. Data management system
The application of an available data management system in the targeted research infrastructure results in a more efficient performance of tools and workflows since it relieves the users from unnecessary uploads and downloads, which could be very time-consuming in data-intensive analyses. Thus, the possibility to directly access the supported data management system via services in the science gateway infrastructure is beneficial for the efficiency of applications. Analogue to job management systems, each data management system possesses its own security features with authentication, authorization and accounting mechanisms.

S3. Data transfer protocols
Additionally to the aforementioned data management systems, research infrastructures provide data transfer protocols for transferring data in general - whether it consists of executables, scripts, small or large input and output data sets or databases. Thus, the science gateway infrastructure needs to support at least one of the available data transfer protocols applied in the research infrastructure to be able to transfer files at all.

The aforementioned challenges are considered a minimal set regarding the technical access to research infrastructures via science gateways.

5 Conclusion

The overall goal of science gateways in HPC is the increased usability of modeling of data and simulations using complex underlying computing infrastructures. This chapter introduces the generic architecture of science gateways and examples for mature solutions. It outlines the importance of gathering information for designing science gateways for domain researchers, who want to apply HPC infrastructures and distributed data management. We have presented checklists for developers in interdisciplinary projects considering domain-related, organizational and technical aspects independent of a specific selected technology but as starting point for selecting technologies and designing as well as implementing a science gateway for a community. These checklists can be used to prepare the interaction with domain researchers and to make informed decisions about technologies suitable for specific science gateways. A wide range of mature and maintained web frameworks and science gateway technologies are available to aid developers in designing and implementing such solutions. While each of them have their own communities, they also have their specific advantages and disadvantages for use cases. Aspects such as scalability and feature availability narrow the scope and help to choose the most suitable technology.

Bridging the differences between research infrastructures via science gateways is a promising way to set the stage for reusability of scientific methodologies and reproducibility of research on an international stage. However, current technical implementations are not sufficient to achieve such goals but also the researchers themselves have to be willing to open up their methodologies and data to the community. Science gateways can be beneficial for this purpose and ease the required steps – especially if they are available in diverse research infrastructures. We have elucidated the challenges faced when science gateways are ported to various research infrastructures in general. While the technical challenges can be summarized in a minimal set consisting of security mechanisms, tool and data availability as well as data management and data transfer protocols, the challenges resulting from policies are dependent on the policies themselves.

Acknowledgements This chapter is based upon work from COST Action IC1406 High-Performance Modelling and Simulation for Big Data Applications (cHiPSet), supported by COST (European Cooperation in Science and Technology).

References

1. Lawrence, K.A., Wilkins-Diehr, N., Wernert, J.A., Pierce, M., Zentner, M., Marru, S.: Who cares about science gateways?: a large-scale survey of community use and needs. In: Proceedings of the 9th Gateway Computing Environments Workshop (GCE '14), pp. 1–4. IEEE Press, Piscataway, NJ, USA. https://doi.org/10.1109/GCE.2014.11 (2014)
2. Dooley, R., Hanlon, M.R.: Recipes 2.0: building for today and tomorrow. Concurrency Computat.: Pract. Exper. **27**, 258 (2015)
3. Liu, J., Pacitti, E., Valduriez, P., Mattoso, M.: A survey of data-intensive scientific workflow management. J. Grid Comput. **13**, 457–493 (2015)
4. IWSG (International Workshop on Science Gateways). http://iwsg.info/
5. Gateway Workshops. http://sciencegateways.org
6. Gesing, S., Wilkins-Diehr, N., Barker, M., Pierantoni, G.: Special issue on science gateways. J. Grid Comput. **14**(4), 495–703 (2016)
7. IEEE Technical Area on Science Gateways. http://ieeesciencegateways.org
8. National Science Foundation (NSF). http://nsf.gov
9. Gesing, S., Wilkins-Diehr, N., Dahan, M., Lawrence, K., Zentner, M., Pierce, M., Hayden, L.B., Marru, S.: Science gateways: the long road to the birth of an institute. In: Proceedings of HICSS-50 (50th Hawaii International Conference on System Sciences), 4–7 Jan 2017, Hilton Waikoloa, HI, USA. http://hdl.handle.net/10125/41919
10. Engelbart, D.C.: Augmenting Human Intellect: A Conceptual Framework, Summary Report AFOSR-3233, Stanford Research Institute, Menlo Park, CA (Oct 1962)
11. The WorldWideWeb Browser. http://www.w3.org/People/Berners-Lee/WorldWideWeb.html (2016). Accessed 29 Feb 2016
12. ISO 9241-1:1992. http://www.iso.org/iso/iso_catalogue/catalogue_ics/catalogue_detail_ics.htm?csnumber=16873 (2016). Accessed 29 Feb 2016
13. AJAX. http://www.w3schools.com/ajax/ (2016). Accessed 29 Feb 2016

14. jQUERY. https://jquery.com/ (2016). Accessed 29 Feb 2016
15. jsPlumb. https://github.com/sporritt/jsPlumb (2016). Accessed 29 Feb 2016
16. WebGL. https://www.khronos.org/news/press/khronos-releases-final-webgl-1.0-specification (2016). Accessed 29 Feb 2016
17. Semantic UI. http://semantic-ui.com/ (2016). Accessed 29 Feb 2016
18. ReactJS. http://reactjs.net/ (2016). Accessed 29 Feb 2016
19. Foundation. http://foundation.zurb.com/ (2016). Accessed 29 Feb 2016
20. AngularJS. https://angularjs.org/ (2016). Accessed 29 Feb 2016
21. Krasner, G.E., Pope, S.T.: A cookbook for using the model-view controller user interface paradigm in Smalltalk-80. J. Object-Oriented Program. **1**(3), 26–49 (1988)
22. Drupal. https://drupal.org/ (2016). Accessed 29 Feb 2016
23. Joomla. http://www.joomla.org/ (2016). Accessed 29 Feb 2016
24. Django. https://www.djangoproject.com/ (2016). Accessed 29 Feb 2016
25. Inc. Liferay. Liferay. http://www.liferay.com (2016). Accessed 29 Feb 2016
26. Apache Software Foundation. Pluto (2016). Accessed 29 Feb 2016
27. Abdelnur, A., Hepper, S.: JSR168: Portlet specification. http://www.jcp.org/en/jsr/detail?id=168 (2003). Accessed 29 Feb 2016
28. Nicklous, M.S., Hepper, S.: JSR 286: Portlet specification 2.0. http://www.jcp.org/en/jsr/detail?id=286 (2008). Accessed 29 Feb 2016
29. Kacsuk, P., Farkas, Z., Kozlovszky, M., Hermann, G., Balasko, A., Karoczkai, K., Marton, I.: WS-PGRADE/gUSE generic DCI gateway framework for a large variety of user communities. J. Grid Comput. **10**, 601–630 (2012)
30. Wolstencroft, K., Haines, R., Fellows, D., Williams, A., Withers, D., Owen, S., Soiland-Reyes, S., Dunlop, I., Nenadic, A., Fisher, P., Bhagat, J., Belhajjame, K., Bacall, F., Hardisty, A., Nieva de la Hidalga, A., Balcazar Vargas, M.P., Sufi, S., Goble, C.: The Taverna workflow suite: designing and executing workflows of Web Services on the desktop, web or in the cloud. Nucleic Acids Res. **41**(W1), W557–W561 (2013). [Online]. Available: http://nar.oxfordjournals.org/content/41/W1/W557.abstract
31. Ludäscher, B., Altintas, I., Berkley, C., Higgins, D., Jaeger, E., Jones, M., Lee, E.A., Tao, J., Zhao, Y.: Scientific workflow management and the Kepler system. Concurrency Comput.: Pract. Experience. **18**(10), 1039–1065 (August 2006). [Online]. Available: https://doi.org/10.1002/cpe.994
32. Berthold, M.R., Cebron, N., Dill, F., Gabriel, T.R., Kötter, T., Meinl, T., Ohl, P., Sieb, C., Thiel, K., Wiswedel, B.: KNIME: The Konstanz Information Miner. Springer, Berlin (2008)
33. Demuth, B., Schuller, B., Holl, S., Daivandy, J., Giesler, A., Huber, V., Sild, S.: The UNICORE rich client: facilitating the automated execution of scientific workflows. In: 2010 IEEE Sixth International Conference on e-Science (e-Science), pp. 238–245 (2010)
34. Streit, A., Bala, P., Beck-Ratzka, A., Benedyczak, K., Bergmann, S., Breu, R., Daivandy, J.M., Demuth, B., Eifer, A., Giesler, A.: UNICORE 6—recent and future advancements. Ann. Telecommun.-annales des Télécommunications **65**, 757–762 (2010)
35. Goecks, J., Nekrutenko, A., Taylor, J., The Galaxy Team: Galaxy: a comprehensive approach for supporting accessible, reproducible, and transparent computational research in the life sciences. Genome Biol. **11**(8), R86 (2010)
36. McLennan, M., Kennell, R.: HUBzero: a platform for dissemination and collaboration in computational science and engineering. Comput. Sci. Eng. **12**(2), 48–52 (2010)
37. Ardizzone, V., et al.: The DECIDE science gateway. J. Grid Comput. **10**, 689–707 (2012). https://doi.org/10.1007/s10723-012-9242-3
38. Deelman, E., Singh, G., Su, M.-H., Blythe, J., Gil, Y., Kesselman, C., Mehta, G., Vahi, K., Berriman, G.B., Good, J., Laity, A., Jacob, J.C., Katz, D.S.: Pegasus: a framework for mapping complex scientific workflows onto distributed systems. Sci. Program. **13**(3), 219–237 (2005)
39. Marru, S., Gunathilake, L., Herath, C., Tangchaisin, P., Pierce, M., Mattmann, C., Singh, R. et al.: Apache airavata: a framework for distributed applications and computational workflows. In: Proceedings of the 2011 ACM workshop on Gateway computing environments, pp. 21–28. ACM, (2011)

40. Dooley, R., et al.: Software-as-a-service: the iPlant foundation API. In: 5th IEEE Workshop on Many-Task Computing on Grids and Supercomputers (MTAGS) (Nov 2012)
41. Dziubecki, P., Grabowski, P., Krysiński, M., Kuczyński, T., Kurowski, K., Szejnfeld, D.: Easy development and integration of science gateways with vine toolkit. J. Grid Comput. **10**(4), 631–645 (2012)
42. Gesing, S., Dooley, R., Pierce, M., Krüger, J., Grunzke, R., Herres-Pawlis, S., Hoffmann, A.: Gathering requirements for advancing simulations in HPC infrastructures via science gateways. Future Gener. Comput. Syst. (accepted)
43. 830-1998 - IEEE Recommended Practice for Software Requirements Specifications. https://standards.ieee.org/findstds/standard/830-1998.html
44. Krüger, J., Grunzke, R., Gesing, S., Breuers, A., Brinkmann, A., de la Garza, L., Kohlbacher, O., Kruse, M., Nagel, W.E., Packschies, L., Müller-Pfefferkorn, R., Schäfer, P., Schärfe, C., Steinke, T., Schlemmer, T., Warzecha, K.D., Zink, A., Herres-Pawlis, S.: The MoSGrid science gateway—a complete solution for molecular simulations. J. Chem. Theor. Comput. **10**(6), 2232–2245 (2014)
45. Plankensteiner, K., Prodan, R., Janetschek, M., Fahringer, T., Montagnat, J., Rogers, D., Harvey, I., Taylor, I., Balaskó, A., Kacsuk, P.: Fine-grain interoperability of scientific workflows in distributed computing infrastructures. J. Grid Comput. **11**, 429 (2013). https://doi.org/10.1007/s10723-013-9261-8
46. SHIWA (SHaring Interoperable Workflows for Large-scale Scientic Simulations on Available DCIs). http://www.shiwa-workflow.eu/project (2016)
47. Gesing, S., Herres-Pawlis, S., Birkenheuer, G., Brinkmann, A., Grunzke, R., Kacsuk, P., Kohlbacher, O., Kozlovszky, M., Krüger, J., Müller-Pfefferkorn, R., Schäfer, P., Steinke, T.: A science gateway getting ready for serving the international molecular simulation community. In: Proceedings of Science, PoS(EGICF12-EMITC2)050 (2012)
48. Zhao, J., Gomez-Perez, J.M., Belhajjame, K., Klyne, G., Garcia-Cuesta, E., Garrido, A., Hettne, K., Roos, M., De Roure, D., Goble, C.: Why workflows break understanding and combating decay in Taverna workflows. In: 2012 IEEE 8th International Conference on E-Science (e-Science), pp. 1–9. IEEE (2012)
49. EGI – European Grid Infrastructure. http://www.egi.eu/ (2016). Accessed 29 Feb 2016
50. XSEDE. https://www.xsede.org/ (2016). Accessed 29 Feb 2016
51. PRACE. http://www.prace-ri.eu/ (2016). Accessed 29 Feb 2016
52. InCommon. https://www.incommon.org/ (2016)
53. Docker. https://www.docker.com/ (2016)

MobEmu: A Framework to Support Decentralized Ad-Hoc Networking

Radu-Ioan Ciobanu, Radu-Corneliu Marin and Ciprian Dobre

Abstract Opportunistic networks (ONs) are an extension of mobile ad hoc networks where nodes are generally human-carried mobile devices like smartphones and tablets, which do not have a global view of the network. They only possess knowledge from the nodes they encounter, so well-defined paths between a source and a destination do not necessarily exist. There are plenty of real-life uses for ONs, including, but not limited to, disaster management, smart cities, floating content, advertising, crowd management, context-aware platforms, distributed social networks, or data offloading and mobile cloud computing. In order to implement and test a routing or dissemination solution for opportunistic networks, simulators are employed. They have the benefit of allowing developers to analyze and tweak their solutions with reduced costs, before deploying them in a working environment. For this reason, in this chapter we present MobEmu, an opportunistic network simulator which can be used to evaluate a user-created routing or dissemination algorithm on a desired mobility trace or synthetic model.

1 Introduction

In the past years, mobile devices (such as smartphones, tablets, or netbooks) have become almost ubiquitous, which has led to the advent of several new types of mobile networks [1]. Such networks are composed almost entirely of mobile devices, and differ considerably from the classic wired networks, both in terms of structure, but also with regard to the protocols and algorithms used for routing and data dissemination. Since there is no stable topology, nodes in mobile networks are not aware of a global structure and have no knowledge of their relationship with other nodes (like

R.-I. Ciobanu (✉) · R.-C. Marin · C. Dobre
University Politehnica of Bucharest, Splaiul Independentei 313,
Bucharest, Romania
e-mail: radu.ciobanu@cs.pub.ro

R.-C. Marin
e-mail: radu.marin@cti.pub.ro

C. Dobre
e-mail: ciprian.dobre@cs.pub.ro

J. Kołodziej et al. (eds.), *Modeling and Simulation in HPC and Cloud Systems*,
Studies in Big Data 36, https://doi.org/10.1007/978-3-319-73767-6_6

proximity, connection quality, etc.). Each node is only aware of information about the nodes that it is in contact with at a certain moment of time, and may act as data provider, receiver, and transmitter, during the time it spends in the network. Thus, a node can produce data, carry them for other nodes and transmit them, or receive them for its own use.

One type of such mobile networks that have been deeply researched in recent years is represented by opportunistic networks (ONs), which are a form of delay-tolerant networks (DTNs). They have evolved naturally from mobile ad hoc networks (MANETs), which store routing information and update frequently. Opportunistic networks are dynamically built when mobile devices collaborate to form communication paths while users are in close proximity. They are based on a store-carry-and-forward paradigm [2], which means that a node that wants to relay a message begins by storing it, then carries it around the network until the carrier encounters the destination or a node that is more likely to bring the data close to the destination, and then finally forwards it. ONs have also gained popularity because they come as an alternative to using the existing wired infrastructures, which may lead to significant power reduction, as well as the decongestion of said infrastructures.

Figure 1 presents an example of the behavior of an opportunistic network. Let us assume that Alice wants to send a message to Bob (using her smartphone), but she does not have access to a wireless infrastructure. Alice composes the message, which is then stored on her device until a contact opportunity arises. Later, Alice goes for a walk and encounters Chris, who also has a mobile device. The opportunistic algorithm decides that Chris is a good relayer, so it sends him the message for Bob (at time t_1). Chris will then continue carrying the message until he encounters Daisy (at time t_2), to whom the message is then relayed further. The cycle can continue on like this, until finally the message arrives at Bob (at time t_3). Thus, it can be seen that data spreading in opportunistic networks employs a probabilistic approach, since no actual paths exist between nodes, and thus no routing tables are present. When a message is relayed to a node, it is not necessarily deleted from the carrier node, so multiple copies of the same message can exist in the network at the same time. The more copies there are, the higher the chance of the message reaching its destination, but flooding the network with too many messages can easily lead to congestion. Opportunistic algorithms employ various methods for increasing the chances of a successful delivery method, while reducing the latency and the network congestion.

There are several challenges regarding the implementation of opportunistic networking in real-life. One of the main challenges is deciding which nodes should the data be relayed to in order for them to reach their destinations efficiently. Various types of solutions have been proposed, ranging from disseminating the information to every encountered node in an epidemic fashion, to selecting the nodes with the highest social coefficient or centrality. Prediction methods have also been employed, based on the knowledge that the mobile nodes from an opportunistic network are devices belonging to humans, which generally have the same movement and interaction patterns that they follow every day. The analysis of contact time (duration of an encounter between two nodes) and inter-contact time (duration between consecutive contacts of the same two nodes) has also been used in choosing a suitable relay node.

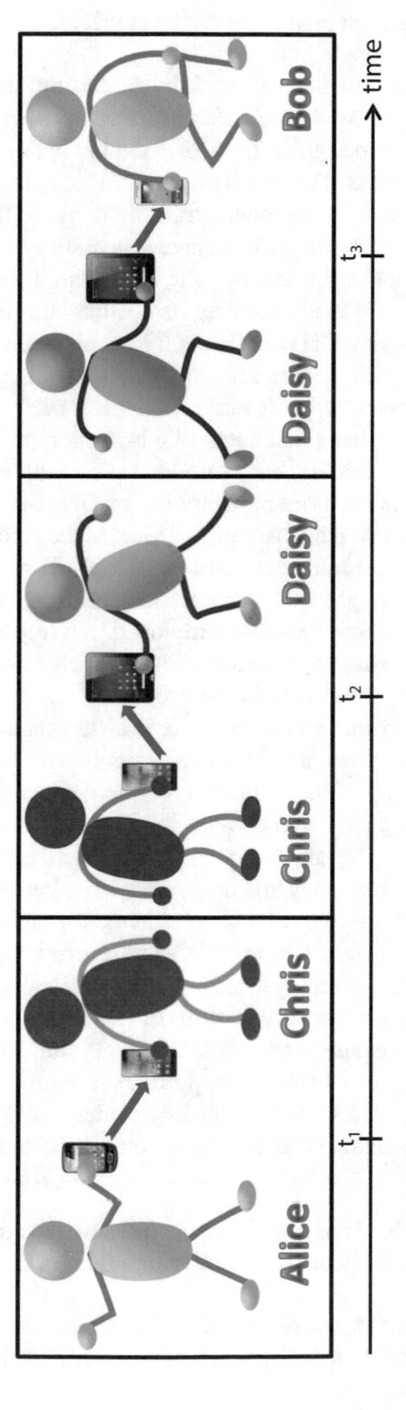

Fig. 1 Opportunistic network interaction

Aside from selecting the node that the data will be forwarded to, research has also focused on congestion control, privacy, security, or incentive methods for convincing users to altruistically participate in the network.

Collaboration between mobile devices implies that the messages sent by such a device can successfully reach their destinations. Moreover, even if ONs are delay-tolerant networks, some applications may require data to be delivered quickly, before they become irrelevant. This is why opportunistic network algorithms should strive to achieve high hit rates, together with low delivery latencies. The hit rate signifies the percentage of messages that have successfully reached their intended destinations, while the delivery latency is the time passed between the generation of a message and the time it is received by its destination. In some ON scenarios, such as disaster management, ON users must have a high degree of confidence that the messages sent reach their destinations, because these are potential life-or-death situations. This is the reason why hit rate is often regarded as the most important metric in ONs. However, other side effects of a high hit rate, such as congestion or high resource usage (CPU, battery, etc.) should also be avoided, if possible.

Aside from this, another important limitation of ONs that should be taken into consideration is that nodes do not have a global view of the entire network, since they are dealing with a fully decentralized approach. Therefore, routing or dissemination algorithms are only able to use information collected from encountered nodes, so mechanisms such as gossiping are often employed. An important direction in ON research deals with increasing an opportunistic network's efficiency, especially with regard to hit rate, delivery latency, and congestion.

Looking at the challenges shown above, it is clear that creating efficient routing and dissemination solutions for ONs is a difficult task. For this reason, frameworks that allow the testing of such solutions before deploying them in real-life are extremely useful. Developers would be able to test their solutions and tweak them in controlled scenarios, being able to change them on the fly, without having to incur high costs (in terms of money and time). However, the challenge in simulating mobile networks arises from two difficult problems: formalizing mobility features and extracting mobility models. Currently, there are two types of mobility models in use: real mobile user traces and synthetic models. Basically, traces are the results of experiments recording the mobility features of users (location, connectivity, etc.), while synthetic models are pure mathematical models which attempt to express the movement of devices. In order to test an opportunistic solution, a simulator that can replay a real-life trace or run a synthetic mobility model, and then apply a given routing or dissemination algorithm, is needed. Thus, our contributions in this chapter are as follows:

- We perform an analysis of the most important existing synthetic mobility models (in Sect. 2) and real-life traces (in Sect. 3), highlighting their benefits and drawbacks.
- We present MobEmu, an opportunistic network simulator, which can run a user-created routing or dissemination algorithm on a desired mobility trace or synthetic model (in Sect. 4).

2 Synthetic Mobility Models

As previously stated, the evaluation of opportunistic solutions can be done in two ways. One way of testing an ON is to use mobility models. Several such models were proposed along the years, ranging from basic random models, to map-based or social-based models. In this section, we present the most relevant mobility models, and highlight the benefits and drawbacks of using such models as opposed to mobility traces collected from real-life situations.

2.1 Random Models

One of the first random models was the Random Walk [3], where nodes move by randomly choosing a direction and a speed, then travel for a predefined time t or distance d, after which a new direction and a new speed are randomly chosen. When a border of the simulation area is reached, the node bounces off depending on the angle of its direction, and continues traveling with the selected speed. Multiple versions of the Random Walk model exist, such as 1-D Random Walk, 2-D Random Walk, etc. A random walk on a one or two-dimensional surface returns to the origin with a probability of 1.0, which ensures that the model tests the movements of entities around their starting points.

The Random Waypoint model [4] is very similar to the Random Walk model, except that, after a node finishes traveling (i.e., time t expires or distance d is walked), it pauses for a predefined time period before choosing a new direction and speed. The Random Direction model [5] is also similar, but nodes must travel until the edge of the simulation area before pausing and then choosing a new direction and speed, instead of using the t or d thresholds. This is done in order to decrease the probability of nodes traveling through the center of the simulation area, which is high for the Random Waypoint model.

Other mathematical models are able to control part of the randomness of existing solutions, such as the Boundless Simulation Area model [6], the Gauss-Markov model [7], or the Probabilistic Random Walk model [8] (other similar models are presented in [9]). However, the disadvantage of random models is that they do not have a good similarity to real-life patterns, where users are grouped into communities and social circles. Thus, real users do not move around randomly, but have very specific destinations that are visited regularly. Moreover, the movements in random models are not realistic, since nodes can have any direction, whereas, in reality, human movement follows streets, walkways, buildings, etc.

2.2 Map-Based Models

Map-based models attempt to make node movements more realistic by selecting points on a map as a node's next position. Thus, for the Random Map-Based model [10], a node's speed is chosen randomly, while the direction is chosen from a set of allowed directions (i.e., that do not pass through walls, the middle of a street, etc.). A node moves until it encounters a restriction (such as a wall), and then a new direction and speed are generated. For the Shortest Path Map-Based model [11], a correct destination is chosen randomly from the valid positions on the map (or from a list of points of interest, or PoIs), and then a shortest path algorithm such as Dijkstra is employed to compute the path that the node must take until it reaches the destination. Upon doing so (with a randomly-chosen speed), the node computes a new destination and speed, and repeats the previous steps. Finally, the Routed Map-Based model [12] allows nodes to have predefined routes, to mimic real-life movement patterns (such as buses, trams, cars, etc.).

The problem with map-based models is that they are still not realistic enough, and are more useful for vehicular networks than for opportunistic networks. Node movement is not governed by social connections and is basically still random.

2.3 Social-Based Models

Social-based models take into consideration the social aspect of human movement, and one such example is CMM [13], which models the degree of social interaction between two people using a value between 0 and 1, and isolates highly connected sets of nodes into social groups based on their centrality. HCMM, or the Home-Cell Mobility Model [14], takes this approach one step further by assuming that nodes in an opportunistic network are not driven only by the social relationships between them, but also by the attraction of physical locations. Thus, each community has a home cell. This mobility model is based on the caveman model [15] and assumes that each node is attracted to its home cell according to the social attraction exerted on that node by all nodes that are part of its community. According to this model, the attraction of an external cell is computed based on the relationships with nodes that have their home in that cell. When a node reaches a cell that is not its own home community cell, it stays there with a probability p_e, and returns to its home cell with the probability $1 - p_e$. An HCMM node starts with a preset community, having strong links with the composing nodes. However, based on a rewiring probability p_r, a node's links can be randomly rewired towards nodes from different communities. Furthermore, a node's decision to leave its current cell or not is taken based on the remaining probability p_{rem}.

The Working Day model [16] attempts to make node movements more realistic by modeling three major activities that humans perform during a weekday: sleeping (at home), working (at the office), and going out with friends (in a restaurant, in the evening). Depending on the time of day, one of these activities is simulated, by also using three separate transport models. Thus, nodes can move (alone or in groups) by walking, driving, or riding a bus, which increases the heterogeneity of movement. Furthermore, the Working Day mobility model also takes into consideration social relationships and communities, which are composed of nodes that either live together, work in the same office, or go to the same pubs in the evening. The advantage of the Working Day model is that the distributions of contact and inter-contact times are similar to the ones found in real-life traces.

2.4 Discussion

The main advantage of mobility models is that they allow the fine-tuning of many parameters, as opposed to mobility traces, which have coarse temporal or spatial resolution and coverage, while possibly exhibiting bias due to an incorrect choice of participants. Although they have been regarded as suspect models due to the limitations in mapping over reality [17], synthetic models have been largely used in the past. However, Barabási [18] introduced a queuing model which disproved the claims of synthetic models based on random walks on graphs. Furthermore, Barabási's model showed that the distributions of inter-event times in human activity are far from being normal, as they present bursts and heavy tails. This happens because people do not move randomly, but their behavior is activity-oriented [19–21]. This endeavor has paved the way for researchers in human dynamics, as the Barabási model [18] is continuously being developed [22–25], and experiments with it are using a variety of new interesting sources: web server logs, cell phone records, or wireless network user traces.

Thus, we believe that random models should not be employed for opportunistic network testing. Instead, more complex models that take advantage of social information, and that manage to obtain contact and inter-contact time distributions similar to real-life traces should be used. Moreover, we also believe that opportunistic algorithms should also be tested on mobility traces, which is why the MobEmu simulator that we propose and present in Sect. 4 offers support for both mobility models and traces.

3 Mobility Traces

In this section, we address mobility traces, i.e., datasets that are collected after performing a tracing experiment where users carry mobile devices that record their interactions and movement patterns.

3.1 Tracing Experiments

This subsection describes the mobility traces that MobEmu currently supports (i.e., that are implemented in the latest version), including two traces that were performed by us at the University Politehnica of Bucharest (UPB 2011 and UPB 2012). At the end of this section, Table 1 shows details regarding all the presented traces.

3.1.1 St. Andrews

The St. Andrews trace [26] was collected using a mobile sensor network with Tmote Invent devices carried by 27 participants from the University of St. Andrews: 22 undergraduate students, 3 postgraduate students, and 2 members of the staff. The experiment was performed for a period of 79 days, in which the participants were asked to carry their devices and to keep them on at all times, whether in or out of the town of St. Andrews.

The Invent devices were able to detect and store information about encounters between each other within a radius of 10 m, and were programmed to send discovery beacons at every 6.67 s. The encounter information, comprised of timestamp, and the scanning and detected devices' IDs, was occasionally uploaded to one of three base stations across the two Computer Science buildings located in the campus of the university. This information was used to create a trace of encounters between Tmotes during the duration of the experiment (the detected social network, or DSN). In addition, a topology (the self-reported social network, or SRSN) was generated using the participants' Facebook information. The nodes were logically split into three large roles according to the SRSN and four weakly-defined roles according to the DSN.

3.1.2 Haggle Traces

Haggle[1] was a European Commission-funded project that designed and developed solutions for opportunistic networks communication, by analyzing all aspects of the main networking functions, such as routing and forwarding, security, data dissemination, and mobility traces and models [27]. The results proposed in Haggle were soon followed by a series of other subsequent research projects targeting similar interests: SCAMPI [28], SOCIALNETS [29], etc. Haggle is today seen by many as the project that created the premises for the advancements on human mobility for information and communications technology-related aspects.

Several mobility traces have been performed in the context of the Haggle project, mostly using Bluetooth-enabled devices such as iMotes. These are mobile devices

[1] http://cordis.europa.eu/project/rcn/80657.

Table 1 Mobility traces statistics

Trace	Devices		Duration (days)	Communication	Trace type	Social data	Interest data
	Mobile	Fixed					
St. Andrews	27	0	79	Bluetooth	Academic and urban	Yes	No
Intel	8	1	6	Bluetooth	Academic	No	No
Cambridge	12	0	5	Bluetooth	Academic	No	No
Infocom	41	0	4	Bluetooth	Conference	No	No
Infocom 2006	78	20	4	Bluetooth	Conference	No	Yes
Content	36	18	25	Bluetooth	Urban	No	No
UPB 2011	22	0	25	Bluetooth	Academic	Yes	No
UPB 2012	66	0	64	Bluetooth and Wi-Fi	Academic	Yes	Yes
Sigcomm 2009	76	0	3	Bluetooth	Conference	Yes	Yes
NUS	22341	0	118	Student schedule	Academic	No	No
GeoLife	182	0	1885	GPS	Urban	No	No
SocialBlueConn	15	0	9	Bluetooth	Academic	Yes	Yes
NCCU	115	0	15	Bluetooth	Academic	No	No

created by Intel, based on the Zeevo TC2001P SoC, with an ARMv7 CPU and Bluetooth support. Two iMote traces, called Intel and Cambridge, have been presented and analyzed in [30]. The Intel trace was recorded for six days in the Intel Research Cambridge Laboratory, having 17 participants from among the researchers and students at the lab. However, data from only 9 iMotes could be collected properly, since the others had hardware issues. The Cambridge trace was taken for five days, at the Computer Lab of the University of Cambridge, having as participants 18 doctoral students from the System Research Group (out of which 12 devices were used for the final trace). For both traces, the iMotes performed five-second scans at every two minutes, and searched for in-range Bluetooth devices. Each contact was represented by a tuple (MAC address, start time, end time). Internal and external contacts were analyzed, where encounters between two devices participating in the experiment were considered internal contacts, while encounters with other devices were external contacts. The authors analyzed the distribution of contact and inter-contact times, as well as the influence of the time of day on encounter opportunities. Regarding inter-contact time, the traces showed that it exhibits an approximate power law shape, which means that inter-contact distribution is heavy-tailed. The authors showed this observation to hold regardless of the time of day, by splitting a day into three-hour time intervals and noticing that the resulting distributions still maintained power law shapes. Contact durations were also noticed to follow power laws, but with much narrower value ranges and higher coefficients.

In addition to the Intel and Cambridge traces, another trace entitled Infocom was presented and analyzed in [31]. It was collected during the IEEE Infocom conference in Miami in 2005, and had 41 conference attendees as participants, for a total duration of four days. The conclusions were similar to the ones above, namely that the distribution of the inter-contact times between two nodes in an opportunistic network is heavy-tailed over a large range of values, and that it can be approximated to a power law with a less than one coefficient. The authors showed that certain mobility models (such as the Random Waypoint model) do not approximate the real-life traces correctly. Similarly to the Infocom trace, another trace (called Infocom 2006) was performed the following year at the same conference (in Barcelona), but on a larger scale. The Infocom 2006 trace [32] was collected between April 24 and April 26 2006. The nodes in the tracing experiment were also iMote devices given to 78 participants at the student workshop, along with 20 static long-range iMotes deployed throughout the workshop area. Interest information about the nodes was collected through questionnaires given to participants, where they were asked to fill in information such as nationality, school, languages, affiliations, positions, city of residence, and topics of interest selected from among those of the workshop's CFP. Thus, the trace contains a total of 18 different topics, with an average of 14.53 topics per node.

Finally, the set of Haggle traces also includes Content, which differs from the traces presented so far, since it was not recorded in an academic or conference environment. Instead, it contains sightings recorded in various locations around the city of Cambridge that are likely to be visited by many people, such as grocery stores, pubs, market places, and shopping centers. The participants in the experiment were students from Cambridge University, but also a series of stationary nodes placed in the key places described above. There were 18 such fixed nodes out of a total of 54.

3.1.3 UPB 2011

The UPB 2011 trace [33] is the result of a social tracing experiment that we performed between November and December 2011 at the University Politehnica of Bucharest, which shows not only the interactions of the experiment participants, but also the social relationships they have with each other according to their Facebook profiles. The experiment collected mobility data using an Android application called Social Tracer. The participants were asked to run the application whenever they were in the faculty grounds. Social Tracer sent regular Bluetooth discovery messages at certain intervals, looking for any type of device that had its Bluetooth on. These included the other participants in the experiment, as well as phones, laptops, or other types of mobile devices in range.

When encountering another Bluetooth device, the Social Tracer application logged data containing its address, name, and timestamp. The address and name were used to uniquely identify devices, and the timestamp was used for gathering contact data. Data logged were stored in the device's memory, therefore every once in a while participants were asked to upload the data collected to a central server located within the faculty premises. All gathered traces were then parsed and merged to obtain a log file. Successive encounters between the same pair of devices within a certain time interval were considered as continuous contacts, also taking into consideration possible loss of packets due to network congestion or low range of Bluetooth. The experiment lasted for a period of 35 days, and involved a total of 22 participants, chosen as statistically varied as possible in order to obtain a good approximation of the mobility aspects of a real academic environment. Thus, there were twelve Bachelor students (one in the first year, nine in the third, and two in the fourth), seven Master students (four in the first year and three in the second), and three research assistants. The participating members were asked to start the application whenever they arrived at the faculty and to turn it off when they left.

It is shown in [33] that the UPB trace follows Hui et al.'s observations [34], namely that the contact and inter-contact times correspond to a power law distribution. It is also shown that the participants have been chosen well so that they represent different groups from the social and logical grouping of nodes in a network based on mobile device carriers in an academic environment. Finally, the k-CLIQUE algorithm [35] has been applied on the trace to show that the local grouping of participants into study years yields similar results to a dynamic grouping such as k-CLIQUE, as well as to the social network organization.

3.1.4 UPB 2012

We collected the UPB 2012 trace [36] at the University Politehnica of Bucharest in the spring of 2012. For this experiment, we implemented an application entitled HYCCUPS Tracer, with the purpose of collecting contextual data from Android smartphones. It was ran in the background and collected availability and mobile interaction information such as usage statistics, user activity, battery statistics, or sensor data, but it also gathered information about a device's encounters with other nodes or with wireless access points. Encounter collection was performed in two ways, Bluetooth and AllJoyn.[2] Bluetooth interaction scanned for paired devices in the immediate vicinity and stored contact information such as the ID of the encountered device and the time and duration of contact. The information stored by AllJoyn tracing was similar, but was collected by constructing and deleting wireless sessions using the AllJoyn framework based on Wi-Fi. The difference between Bluetooth and Wi-Fi is that Wi-Fi consumes more battery life, but is more stable. We observed that AllJoyn interactions occurred much more often than those on Bluetooth. Thus, there were 20,658 Wi-Fi encounters for a total of 66.27% of all the contacts, and only 6969 Bluetooth contacts. We believe that such results were caused by the low range of Bluetooth. Tracing was executed periodically with a predefined timeout for Bluetooth, and asynchronously for AllJoyn interactions.

The duration of the tracing experiment was 64 days, between March and May 2012, and had 66 participants. They were chosen so that they covered as many years as possible from both Bachelor and Master courses. Thus, there were one first-year Bachelor student, one third-year Bachelor student, 53 fourth-year Bachelor students (from five different study directions), three Master students, two faculty members, and six external participants (from an office environment). We were interested only in the participants at the faculty, so we eliminated the external nodes. We also eliminated some nodes that had too little contact information, because they were irrelevant to our experiment. Such nodes belonged to students that did not keep their Android application on at all times when they were at the faculty as they were instructed, or who did not attend many classes in the experiment period. In the end, we remained with 53 nodes with useful information in the trace. By analyzing the participants' Facebook profiles, we were also able to extract the social connections matrix, as well as the users' interests. There were five global topics, each participant having in average 3.51 interests.

3.1.5 Other Traces

The Sigcomm 2009 trace [37] was collected using an opportunistic mobile social application entitled MobiClique. The tracing experiment lasted for three days and gathered data from 76 smartphones running MobiClique, which were given to participants of the Sigcomm 2009 conference in Barcelona. When receiving the

[2] https://allseenalliance.org/framework.

smartphones, volunteers were asked to fill in their interest data using the Mobi-Clique application, which were then exported anonymously for the trace. There are 151 total topics in the trace, with an average of 15.61 topics per node. NUS [38] is a dataset of contact patterns among students, collected during the spring semester of 2006 at the National University of Singapore, while GeoLife[3] is a GPS trajectory dataset collected from 182 users in a period of over three years. SocialBlueConn [39] contains traces of Bluetooth encounters, Facebook friendships, and interests of a set of users, collected through the SocialBlueConn application at University of Calabria and, finally, NCCU [40] contains contact information for a group of college students in a campus environment.

3.2 *Discussion*

The main reason for developing and using tracing applications such as the ones previously presented (instead of synthetic mobility models) spawns from the need for better mapping onto real-life situations. As previously stated, trace models follow a heavy-tailed distribution with spikes and bursts, making random models obsolete.

The major benefit of tracing applications is the use of a custom data model in order to relate to real situations, real problems, and optimized solutions for said issues. However, this can also lead to a pitfall: if the data model is not correctly designed at the start of the experiment, the entire outcome of the analysis can be biased.

Among the potential challenges of setting up our tracing experiments (UPB 2011 and UPB 2012), we dealt with the following:

- Finding volunteers representative to our goals was not such an easy task as it may seem. For example, if we would have chosen all participants from the same class, then our results would have been biased because we would have been limiting our targeted scope to a partition of our community graph instead of reaching the entire collective. Moreover, all of the candidates for the experiment needed to have Android devices capable of tracing our data model: Bluetooth connectivity, Wi-Fi connectivity, sensors etc.
- The design and development of the tracing applications needed to take into account compatibility with multiple types of viable Android devices of varied versions. Furthermore, when developing the tracers, we were obliged to take into account the additional overhead of our applications, as most participants complained about the supplementary power consumption.
- The installation effort of the tracer was high due to issues such as Bluetooth pairing: all of the participants' devices needed to pair to each other in order for us to be able to trace their interactions.

[3]https://www.microsoft.com/en-us/download/details.aspx?id=52367.

- Last, but not least, we were confronted with the human factor of such experiments: the lack of conscientiousness of our volunteers. Due to the participants not running the tracing application as instructed, the collected data were incomplete. Furthermore, this affected the analysis of said results, as we needed to deploy measures to deal with uncertainty.

In conclusion, we believe that human mobility traces offer a better representation of real-life opportunistic interactions than random synthetic models. However, great care has to be taken when performing the tracing experiments, in order for the collected data to not be biased. Most importantly, the experiment participants should be incentivized to follow the rules of the experiment and carry their mobile devices with the tracing application started whenever it is necessary. Moreover, in order for the collected trace to represent a correct approximation of the entire environment, the participants have to be chosen so that each social community is represented. A caveat of mobility traces is that they generally have a limited time granularity, since scanning for in-range devices is performed periodically, in order to consume less power. Thus, some contacts may be missed, leading to an incomplete trace. Another disadvantage is that the trace results cannot be scaled, since the number of participants is fixed.

As a conclusion, we recommend employing both real-life traces, as well as social-based mobility models, when testing an opportunistic algorithm. For this reason, the MobEmu simulator that we propose and present in Sect. 4 offers support for both.

4 MobEmu

In order to be able to run and test opportunistic networking solutions on mobility traces or models, a simulator for realistic ON evaluation was required. Since the existing solutions did not offer everything we needed (as shown in Sect. 4.5), we implemented MobEmu, an opportunistic framework used for replaying mobility traces and emulating data routing and dissemination algorithms, which we present in this section.

Since opportunistic networks may be composed of hundreds or thousands of devices (if not more), testing new ideas for data routing and dissemination can prove to be a challenge. Furthermore, if the algorithms do not function as expected on the first go, re-deploying a new version would be necessary, which might prove costly, both in terms of money, as well as time and organizational effort. For this reason, simulators are used, which are able to play back existing traces (collected from members of opportunistic networks) or run synthetic mobility models, so that the creator of an algorithm can have an idea regarding how the algorithm behaves prior to actually releasing it in the network.

MobEmu is such a simulator, which can run a user-created algorithm on a desired mobility trace or synthetic model, as long as certain implementation rules are followed. It is written in Java, so it is highly modular and easy to understand and/or modify, and its source code is freely available on GitHub.[4]

4.1 Functionality

MobEmu's functionality is relatively straightforward. It parses a mobility trace and, at every step of the trace (given by the time unit the trace was measured in), it checks whether a contact between two nodes occurs. If this is the case, then the desired routing or dissemination algorithm is applied for each node, in regard to the encountered node. Moreover, various statistics are collected, as will be shown in the next subsection. Aside from checking for contacts, MobEmu checks at every tick whether a node should generate messages for other nodes, an action which is controlled by the user. Thus, the amount of messages sent, their destinations, priorities, number of copies, TTL, etc., can be configured according to each user's desire. At each step, the emulator also computes a node's community according to the k-CLIQUE algorithm, as well as its local and global centralities (i.e., inside its community, and outside of it).

Regarding the execution of a routing or dissemination algorithm, the user is able to control the data memory size of each node, the amount of history it can store, the speed with which messages can be exchanged, a node's level of altruism, etc. When a trace run is completed, the desired statistics are printed.

4.2 Components

This subsection presents the main components of MobEmu: the *Trace* (along with the *Parser, Context,* and *Contact* components) and the *Node* (containing information regarding *Altruism, Battery, Messages, Network,* etc.). Each of the subcomponents will be described in detail, for a full understanding of MobEmu's purpose and functionality, and they are also shown in Fig. 2.

4.2.1 Trace

The first step in running an opportunistic algorithm is deciding what type of network it will be run in. As we have previously stated, using mobility traces is a cheaper alternative to deploying and testing an algorithm in a real-life network. MobEmu's *Trace* component thus contains a list of *Contacts*, as well as information regarding the trace's name. Since all kinds of mobility traces have been collected at faculties or companies all across the Globe, a trace's sample time may differ, as well as its

[4]https://github.com/raduciobanu/mobemu.

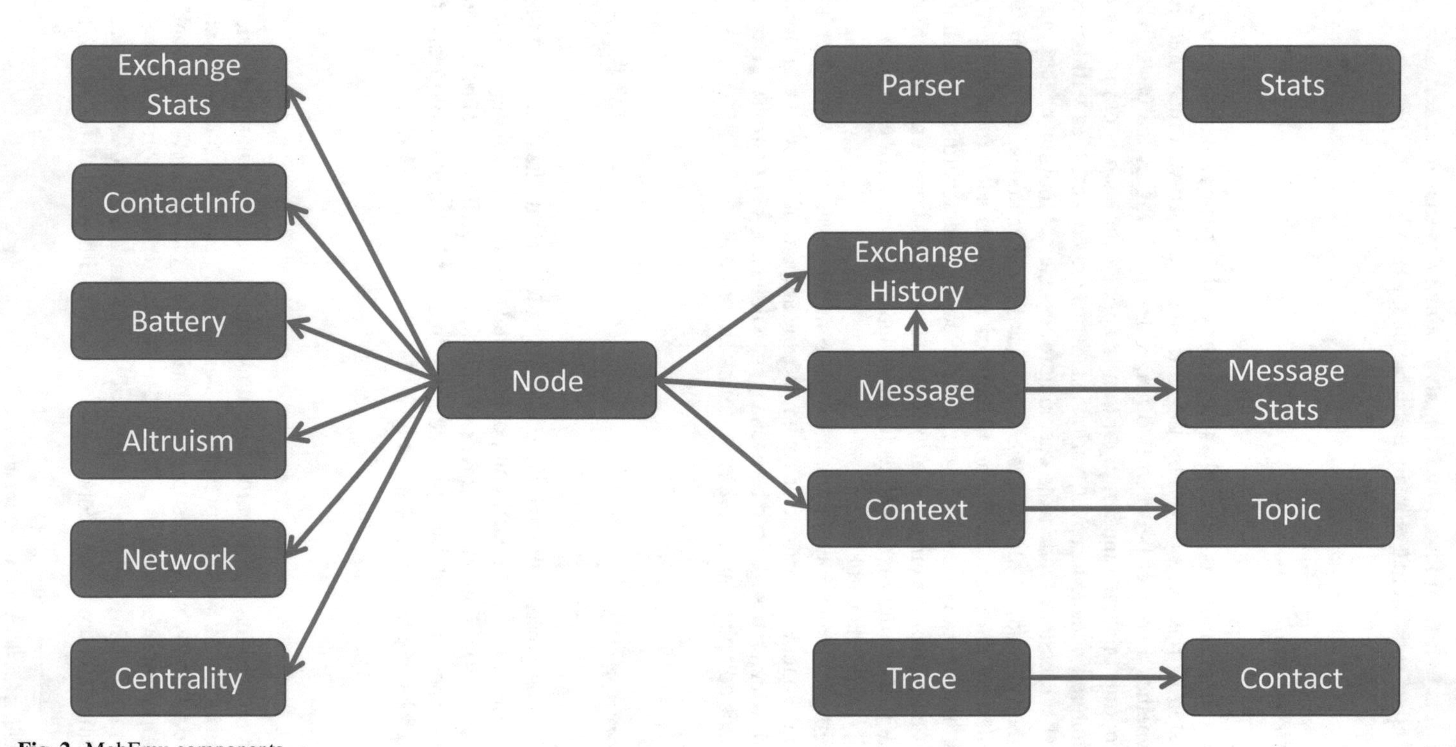

Fig. 2 MobEmu components

starting point (for example, some traces start with timestamp 0, while others use the Linux-style epoch). In order to account for such differences, a *Trace* object also contains information about the trace's start point, end point, and sample time.

Since we stated that a *Trace* contains a list of *Contacts*, it should be noted that such a *Contact* contains information about the two nodes that are in contact, as well as the starting and finishing time of the contact. Some traces contain contacts collected through various means (such as Bluetooth vs. Wi-Fi), so a *Contact* also specifies the type of contact between the nodes. The observer and observed nodes are marked separately, because some traces are not symmetrical, i.e., it is possible that a node A sees a node B, but node B does not see node A, because data are collected by polling and the contact is too short for B to start polling again. On the other hand, even if a contact is seen by both nodes, it is highly possible that their clocks are not in sync, or that the polling was performed at different times. However, MobEmu solves this problem by merging contacts between two nodes, as follows: if a node A sees a contact with node B and, while that contact is still ongoing, B sees a contact with A, the second contact is not considered as a separate contact. In conclusion, a *Trace* contains all the necessary information regarding the contacts between nodes.

MobEmu *Trace* objects are obtained by running a *Parser*, which is an interface that contains the following methods: *getTraceData* (for obtaining the *Trace* object), *getContextData* (for getting context information regarding nodes' preferences), *get-SocialNetwork* (used to obtain the social connections between nodes on online networks such as Facebook or Twitter), and *getNodesNumber*. The last function returns the number of nodes, but is also useful for knowing the IDs of all the nodes in the trace, since they should be numbered consecutively from 0. The next subsection shows how the *Parser* interface should be implemented, in order to correctly obtain a trace that can be used in MobEmu.

As stated above, the result of parsing a trace contains not only the *Trace* object itself, but also a *Context* map, with a *Context* object for each node in the trace. A node's context represents information regarding the node's interests. Thus, a *Context* object contains the node's ID and a set of *Topics* that the node is interested in. These *Topics* are represented as integers, and correspond to real-life interests of the user that the opportunistic node belongs to (such as music, movies, books, etc.).

4.2.2 Node

A *Node* object contains all the information that an opportunistic node requires for running a data routing or dissemination algorithm in MobEmu. Firstly, it contains each node's ID, which is unique in the network, all nodes being consecutive integers starting from 0.

In order to store messages that are carried opportunistically, a node has a data memory. It contains messages received from other nodes and intended for destinations other than the node itself (since information required by the node is directly sent to the application level, when dealing with data routing). Thus, a node's data memory is the means through which the store-carry-and-forward paradigm is enforced, along with the routing or dissemination algorithm. The higher a node's data memory size, the more messages it can carry for other nodes, which offers a great chance of increasing the network's hit rate. However, having many messages in a data memory leads to longer times required for analyzing the messages and deciding which ones should be forwarded upon an opportunistic contact. Aside from the data memory, a node also has a separate memory where it stores the messages it generates itself. They are kept separate, since they are being handled differently, as a node's own messages are never deleted from memory. On the other hand, when the data memory is full, some messages may need to be deleted to make room for the new ones. This is implemented using a cache replacement policy.

Both memories (the data memory and a node's own messages) are represented as lists of *Messages*. Such an object has a unique ID (to separate between messages more easily), the IDs of the source and destination (or, in case we are dealing with dissemination, the ID of the source and a *Context* object representing the tags this message is marked with), as well as a timestamp specifying when it was generated. A string for the actual message text is also a part of the *Message* object. Statistics regarding a message's path through the network are also kept inside the *Message* object, in order to make it easier to track. These stats (implemented using a *MessageStats* object) include whether the message was delivered and to whom, the number of hops the message has traveled, the latency of arriving to the intended destination (or destinations, if performing dissemination), as well as the number of message copies available.

Various opportunistic algorithms need to analyze a node's memory, in order to decide which messages maximize the overall network hit rate. Thus, MobEmu nodes also contain a map of encountered nodes and information about them, represented as *ContactInfo* objects. There is such an object for each encountered node, and it specifies the total duration of contacts with that node (i.e., the sum of the durations of all contacts with that node), the number of contacts, and the last encounter time. This information is updated at every contact between two nodes, and may be used for efficient routing or disseminating.

A node also stores statistics about data exchanges that occur at a contact, through an *ExchangeStats* object, which specifies the time and duration of the last exchange with each encountered node. This is used for aggregating contacts that are seen differently by two nodes (as was described at the beginning of this section). Among other statistics stored by opportunistic network nodes, we would also highlight the number of overflow events (i.e., when the data memory is full and a message is to be received), the total number of messages received, delivered, or exchanged, as well as the number of encounters with every node. The history of data exchanges between nodes is also stored, since it is useful for some algorithms. An *ExchangeHistory* object is used for this purpose, which stores the timestamp of the exchange, information about the message exchanged (such as source and destination), the nodes that performed the exchange, and the battery level of the forwarding node.

We previously specified that, aside from applying the data routing or dissemination algorithm at each step of the trace, MobEmu also performs community detection using the *k*-CLIQUE algorithm. This information is stored in a list of node IDs entitled *localCommunity*. Nodes from this list belong to the current node's local community, as computed by k-CLIQUE based on the *contactThreshold* and *communityThreshold* parameters. As a result of applying k-CLIQUE and detecting communities, each node also has a centrality within that community, as well as a global centrality in the entire network. These values are stored using *Centrality* objects, since information is required about previous and cumulated values as well. A different type of community, the social network, is also stored as an array of boolean values that specify whether there are friendship relationships between the node and other participants in the network.

Since opportunistic network nodes are generally mobile devices that have limited battery life, it is necessary to be able to model the behavior of the MobEmu nodes as if they were such devices. This is done using a *Battery* object, which offers information about a node's current battery level, whether it is recharging or not (and the time left to recharge, if it is), as well as the battery drain model. Consequently, this class contains a method named *updateBatteryLevel*, called at every step of the trace. This method contains for now a linear formula for draining battery, but this is not necessarily realistic. However, the method can be easily overridden in inherited classes, if a different behavior is expected. A node's current battery level dictates whether it is actively participating in tracing or not, since a node with a low battery may choose to attempt to maximize whatever is left, and thus avoid acting as a relay for other nodes. This is enforced using a *minBatteryThreshold* in the *Battery* object of each node.

As we previously stated, we wish to have a node that behaves as close to real-life as possible. Thus, because opportunistic contacts have limited durations, only a certain amount of data can be exchanged at each contact, so each node should have network information. This is implemented through a *Network* object that contains the transfer speed of a node, measured in messages per time unit (assuming all messages have the same size). Consequently, depending on the duration of a contact, nodes will only be able to exchange a certain number of messages (we assume that communication is bidirectional, a node's transfer speed specifying the speed it is receiving data with).

Finally, each node has the context information previously described, represented by a *Context* object, as well as altruism information in the shape of an *Altruism* object. Nodes in opportunistic networks may not always act altruistically, for various causes: low level of battery or other resources (CPU, memory, etc.), lack of interest in helping other nodes (such as non-community nodes), etc. Nodes' refusals to follow the store-carry-and-forward paradigm of opportunistic networks leads to a high decrease of the network hit rate, and an increase of its latency. This not only affects other nodes, but the selfish nodes as well, since they may no longer be helped by nodes that consider them to be selfish. An *Altruism* object in MobEmu contains the node's local and global altruism levels, since they can be different based on social or interest communities. For example, a node may only want to relay data for nodes with common interests, or for social neighbors. Perceived altruism values for the other nodes in the network are also stored in the *Altruism* object (used by selfishness detection and incentive algorithms), as well as a boolean that specifies whether the current node is considered selfish by others or not. When a node is considered selfish, it will not be helped by other devices in the network anymore, so it needs to become more altruistic if it wishes to have its messages delivered.

4.3 Implementing a Mobility Trace Parser

In order to implement a new mobility trace parser in MobEmu, the first step is to implement the *Parser* interface, which was described above. The four functions previously presented should be implemented: *getTraceData* should return a *Trace* object containing the *Contacts* between the nodes, *getContextData* should return a map with integer keys (the IDs of the nodes) and *Context* values (each node's context, which can even contain no tags, if the node did not specify interests when the trace was collected), *getSocialNetwork* should return a symmetrical bidimensional array of boolean values (where an entry specifies whether there is a social connection between two nodes), and *getNodesNumber* should return the total number of nodes in the trace. An important note to be made is that node IDs should be integer values between 0 and $N - 1$ (where N is the value returned by *getNodesNumber*). Parsers for all the traces shown in Table 1 are implemented in MobEmu.

Aside from working with mobility traces, MobEmu can also replay data generated by a mobility model, and the functionality is exactly the same as when replaying a trace. The movement and interaction data are generated by the model and stored as *Contact* objects in a *Trace*, and then the next steps are exactly the same as when dealing with traces. The only difference is that the parameters of mobility models can be configured, ranging from number of nodes to the size of the network. MobEmu currently supports the HCMM model presented in Sect. 2.

4.4 Implementing a Routing or Dissemination Algorithm

The most important step in implementing an opportunistic routing or dissemination algorithm in MobEmu is writing the data exchange function. However, prior to presenting how this is done, we start by describing the steps taken by the emulator when two nodes are in contact.

When two MobEmu nodes meet opportunistically (based on the data contained in the trace), the *exchangeData* function is called by the observer node. This function receives the observed node as a parameter, as well as the duration of the contact and the current trace time. If a contact between the two nodes is already in progress (caused, as specified before, by the differences in clocks or sampling intervals of the nodes), then the function simply returns. However, if the contact just began, the *onDataExchange* function is called. This is an abstract function that should be implemented by any class that extends *Node*. All such classes implement the actual dissemination or routing algorithm through the *onDataExchange* function, while at the same having access to all the information of the current node (which is why we decided to use inheritance, rather than composition).

Thus, all that is required for writing a routing or dissemination algorithm is implementing the *onDataExchange* function, but there are several guidelines that should be followed, as shown in Fig. 3. Firstly, since the first parameter of the *onDataExchange* function is of type *Node*, it should be checked if it is indeed an instance of the implementing class, and cast to it. Then, the next step is generally to perform the delivery of direct messages, which are messages destined for the current node (when dealing with routing) or marked with tags that the node is interested in (if disseminating data). This is done by calling the *deliverDirectMessages* method. Then, the messages from an encountered node's data memory and own memory (i.e., the messages it generated itself) are analyzed and, based on an algorithm-specific decision, they are downloaded by the current node or not. A message is downloaded by calling the *insertMessage* function at the current node. If the network bandwidth is considered restricted, then the user must ensure that more messages than allowed per contact are not downloaded. Altruism can also be taken into account when performing data exchanges upon a contact, but it is the programmer's job to decide whether a node is selfish or not, and whether it should not be helped by other nodes.

Statistics are collected automatically, so the user does not need to do this explicitly, unless more stats are needed. For this, new fields should be added to the *Node* implementation, and they should be updated when necessary, by extending the corresponding functions. Furthermore, the statistics are aggregated by using the *Stats*

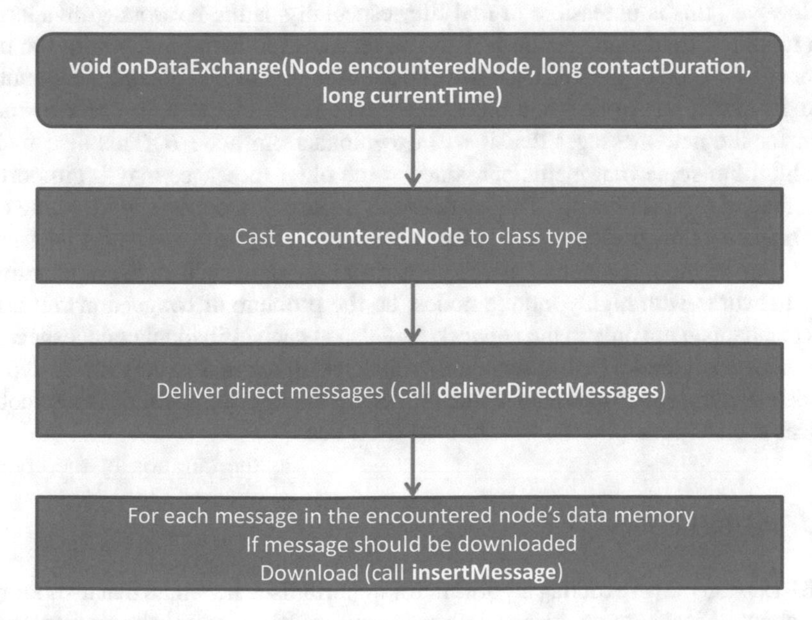

Fig. 3 Steps for implementing a routing or dissemination solution

class, which should be extended, and methods for various statistics should be added as desired.

A user may also wish to change the way messages are generated. Currently, this is done daily, in a time interval around lunch (since this is generally when the most contacts occur for the majority of the traces). If the current algorithm performs point-to-point routing, then messages are generated by default using a Zipf distribution with exponent 1, where the most messages are sent to nodes that are both in a node's community, as well as in its online social network, and the fewest to nodes that have no connection to the current node. If dissemination is performed, a node chooses a random interest, and generates a message marked with it. The *generateMessages* function from the *Message* class can be overridden if a new message generation behavior is desired. The opportunistic routing or dissemination solutions currently implemented in MobEmu are briefly described in the following subsections.

4.4.1 Epidemic

The Epidemic algorithm [41] is based on the way a virus spreads: when two potential carriers meet, the one with the virus infects the other one, if it is not already infected. Thus, when an ON node A encounters a node B, it downloads all the messages from B that it does not already contain, and vice versa. The simplest version of this algorithm assumes that a node's data memory is unlimited, so that it can store all the messages that can be at once in the opportunistic network.

However, this is unfeasible in real-life, especially as the network grows larger, so a modified Epidemic version is also implemented in MobEmu, where the data memory of a node is limited. Thus, when node A's memory is full and it encounters node B, first it has to drop the oldest messages in its memory, in order to make room for the new messages that it will download from node B. This also makes the algorithm somewhat inefficient, since some older messages may be important (e.g., they may be addressed to nodes that A is about to encounter), while some new ones may be totally irrelevant (e.g., their destinations may be nodes that A will never meet). Furthermore, Epidemic nodes perform many data exchanges, especially in large networks with highly mobile nodes, so the problem of congestion can arise. This can happen not only in the network, but also at each individual node, especially if the network is dense. Performing data exchanges often can also very easily deplete a node's battery, since opportunistic networks are mainly composed of small mobile devices such as smartphones, with limited lifetimes.

4.4.2 BUBBLE Rap

BUBBLE Rap [32] is a routing algorithm for opportunistic networks that uses knowledge about nodes' communities to deliver messages. It assumes that a mobile device carrier's role in the society is also true in the network, thus the first part of the algorithm is to forward data to nodes that are more popular than the current node. The second assumption made by BUBBLE Rap is that the communities people form in their social lives are also observed in the network layer, therefore the second part of the algorithm is to identify the members of the destination community and pass them the relevant messages. Thus, a message is bubbled up the hierarchical ranking tree using a global popularity level, until it reaches a node that is in the same community as the destination. Then, the message is bubbled up using a local ranking until it reaches its target. The popularity of a node is given by its betweenness centrality, which is the number of times a node is on the shortest path between two other nodes in the network. Community detection is done using k-CLIQUE, while the centralities are computed by replaying the last collected mobility trace, applying a flooding algorithm, and then computing the number of times a node acts as a relay on a shortest path.

However, this implementation of BUBBLE Rap is unfeasible in real-life, because it has to know the behavior of the nodes beforehand. Therefore, a distributed version entitled DiBuBB is also proposed by the authors [32]. It uses distributed k-CLIQUE for community detection, together with a cumulative or single window algorithm for distributed centrality computation. The single window (S-window) algorithm computes centrality as the number of encounters the current node has had in the last time window (chosen usually to be six hours), while the cumulative window (C-window) algorithm counts the number of individual nodes encountered for each time window, and then performs an exponential smoothing on the cumulated values.

4.4.3 ML-SOR

Whereas solutions like BUBBLE Rap only employ social network information, ML-SOR (Multi-Layer SOcial network-based Routing) [42] also uses interest information, as well as encounter history. Thus, it exploits three social network layers: the online social network, the interest network, and the contact network. The latter is the proximity graph created through contacts between devices, while the online social network is extracted from virtual contacts. Thus, a layer is represented as a weighted graph, where the edges are social links between the nodes, which are the vertices. A tuple of multiple such social network layers is defined as a multi-layer social network.

ML-SOR extracts social network information from multiple contexts and analyzes encountered nodes in terms of node centrality, tie strength, and link prediction, on different social network layers. When an ML-SOR node A encounters a node B, it computes a social metric called MLS for all the messages in B's memory, both from its own standpoint, as well as from B's. If MLS is higher for node A in terms of a message M, then A sends a download request for M. The social metric is computed based on three components: CS, TSS, and LPS. CS represents the centrality of the nodes in the contact history graph, while TSS and LPS are computed with regard to the message's destination. TSS is the online social network strength between the analyzed node and the destination, and LPS is a link predictor computed on an interest network layer. It counts the number of common interests between the encountered node and the message's destination.

Thus, the ML-SOR algorithm tries to forward messages to nodes that are more important than the carrier node on three levels. An important node is defined as a node with a high centrality (i.e., with many encounters), which is connected to the destination on the online social network, and has multiple interests in common with the destination. ML-SOR is therefore based on the assumptions that nodes that are socially connected tend to meet each other more often, and so do nodes with common interests.

4.4.4 Moghadam-Schulzrinne

The dissemination solution proposed by Moghadam and Schulzrinne [43] uses interest information when distributing data. This interest-aware algorithm is able to analyze a user's history of cached data in order to obtain his interests. Using these interests, the algorithm is able to decide whether a document carried by an encountered node should be downloaded or not when a contact occurs. Interests are represented as vectors of interest, which are obtained by applying extended latent semantic analysis and singular value decomposition on the documents that have been viewed or cached on each user's device. When a node A meets a node B, the former has to decide which documents should be forwarded to the latter. This is done by mapping each document carried by A into B's interest space and applying cosine similarity. If the result is higher than a predefined threshold, then that document is transferred to B.

Thus, documents are only spread to nodes that are interested in their content, which in turn can forward them further on to nodes with similar interests. By using such an algorithm, the number of data exchanges in the network decreases dramatically, since a node can only carry a document that it is interested in (and that the nodes it will encounter are, with a high probability, interested in too). The results show that the interest-based algorithm is able to deliver 30% more relevant documents and 35% fewer irrelevant documents, when compared to Epidemic routing.

4.4.5 Social Trust

Social Trust [44] is a trust method that leverages social information to establish trustworthy communication for mobile opportunistic networks. Nodes' trust is social-based, since it is argued that they belong to an opportunistic network composed of people's devices (such as smartphones). Thus, socially-connected nodes have an intrinsic trust in each other, since they are likely to interact more often in good conditions.

The authors propose employing two major techniques of establishing trust: Relay-to-Relay and Source-to-Relay. When using the former method, a node that is carrying a message computes the trust in an encountered peer based on the relationship between the two nodes, while the latter method assumes that candidate relays are analyzed based on their relationship with the message's source. For each of the two trust methods, four ways of computing a node's trust are proposed (and implemented in MobEmu): common interests, common friends, social graph distance, and a combination between common friends and social distance. These filters are based on research in the area of human mobility, which shows that socially-connected people are more likely to encounter each other than to have contacts with stranger nodes. The same goes for interests: people with common interests tend to meet often and on a regular basis. Since opportunistic network nodes are very likely humans carrying mobile devices, it is natural to employ knowledge about human mobility and interactions.

4.4.6 JDER

JDER [45] is an opportunistic routing solution that is based on the idea that there are circumstances when information about social connections is not available, or the social network is much too large to be used efficiently. In such situations, socially-aware algorithms do not behave efficiently, since they are mostly focused on forwarding messages to popular nodes that are likely to communicate with many other nodes in the network. However, there are critical nodes (cut nodes) with less apparent popularity that play an important role in the dissemination process of messages in the network, so they have to be found and selected as forwarders in order to guarantee a high delivery ratio.

The JDER algorithm attempts to find the cut nodes by employing two metrics: the history encountered ratio and the Jaccard distance. The former specifies how often each encountered node has been met, whereas the latter is a measure of similarity, and is computed as the number of common neighbors between two nodes. Thus, if two peers are similar according to the Jaccard distance, then they need to exchange data. Similarly, if an encountered node is similar to the destination of a message, that message must be transferred to the encountered node, which has a high chance of encountering the destination.

4.4.7 IRONMAN

IRONMAN [46] is a selfish node detection and incentive mechanism for opportunistic networks that uses pre-existing social information to detect and punish selfish nodes, incentivizing them to participate in the network. Each IRONMAN node stores a perceived altruism (or trust) value for other nodes, that is initialized based on the social network layout: if the nodes are socially connected, this value is higher than for non-community nodes. When a node A meets a node B, it checks its encounter history to see if B has ever created a message for A that has been relayed to another node C. If this is the case, and A has encountered C after B had given it the message but A did not receive the message, then C is considered selfish, and A's perceived altruism of C is decreased. Whenever a node A receives a message from a node B which is not the source of the message, A's perceived altruism of B is increased.

Apart from detecting selfish nodes, IRONMAN also uses incentives to make nodes behave better. Therefore, whenever a node B is considered selfish by A (its perceived altruism is below a given threshold), it is notified, and A will not accept any messages from it (but will keep on forwarding its own messages to B). Therefore, a selfish node might end up not being able to send its messages, unless it becomes altruistic. IRONMAN uses perceived altruism ratings for encountered nodes, in order to decide if they are selfish or not. These ratings are computed locally based on the analysis of the history of contacts whenever two nodes meet, but the local values are exchanged with other nodes at every encounter, in order to inform them if a node is selfish.

4.4.8 SENSE

SENSE [47] is a novel social-based collaborative content and context-based selfish node detection algorithm with an incentive mechanism, which aims to reduce the issues caused by having selfish nodes in an ON. It uses gossiping and context information to make informed decisions regarding the altruism of nodes in the network, on one hand, and incentive mechanisms to make selfish nodes become altruistic, on the other. SENSE takes advantage of social relationship knowledge regarding the nodes in the ON, to decide if a node is selfish towards its own community.

When a SENSE node A encounters another peer B, A verifies B's reputation before deciding to employ it as a relay. This reputation is computed based on past information regarding B's behavior, obtained not only from A's observations, but also through gossiping from the other nodes in the network. If node A decides that B is altruistic, then it continues to communicate with it normally. However, if node B is considered selfish, then A will not only stop sending its messages to B (i.e., it will not use it as a relay), but it will not help B deliver its messages. Thus, node B will be forced to become unselfish if it wants its messages delivered. The SENSE algorithm uses social information for differentiating between community altruism and selfishness, while also taking into account other context information such as battery level.

4.4.9 SPRINT

SPRINT [48, 49] is a data routing algorithm for opportunistic networks that uses social information and contact prediction to perform forwarding decisions. More specifically, upon a contact between two nodes in an ON, it employs a utility function for computing the value of each message carried by the two nodes. The goal is for each node to maximize the utility of the messages it is carrying, through data exchanges. Thus, each node computes the utility of every message, and then stores the N most valuable ones (where N is the maximum number of messages that a node can transport).

The utility function has two components, the first of them taking into account a message's freshness (i.e., how much time has passed since it has been generated), as well as the probability that the current node is able to bring the message closer to its intended destination. This probability is computed by analyzing a node's contact history and its social connections, and by using a Poisson distribution. The second component of the utility function is based on a node's social connection with the message's destination, the number of hops that the message has traveled through, the node's popularity, and the time spent by the node in contact with the message's destination.

MobEmu also supports a version of SPRINT entitled SPRINT-SELF [50], which contains SENSE mechanisms for trust and reputation when performing opportunistic routing, as well as incentivizing nodes to participate in the ON.

4.4.10 ONSIDE

ONSIDE [51, 52] is an algorithm that not only uses interest knowledge for opportunistic data dissemination, but also social information about the nodes in the network, in an attempt to decrease an ON's overall bandwidth consumption and reduce its congestion, while not affecting the average per-topic hit rate and the delivery latency.

The ONSIDE algorithm is based on several assumptions, one of them being that nodes with common interests tend to meet each other more often than nodes without. The second assumption that ONSIDE is based on states that connections from online social networks (such as Facebook, Google+, or LinkedIn) are respected in an ON node's encounters. Whenever two nodes running ONSIDE meet, they exchange lists of messages in their data memory and lists of topics each node is interested in. Based on this information, each node analyzes the other node's messages and decides which of them should be downloaded. This decision is firstly based on common interests between the encountering nodes, so that data transfers are only performed between nodes with at least one common interest. The second component of the decision function ensures that a node will not only download a message for itself and then drop it after use, but will also store it for others, since it is highly likely to encounter other nodes that have interests similar to its own. The data exchange decision is also based on the idea that a node encounters its social connections often, so, if it has friends interested in a message it encounters, it should download that message in order to ensure a quick delivery. The last component of the exchange function is computed based on the nodes' encounter history, assuming that a node's behavior in an ON is predictable, so that if it encountered many nodes subscribed to a certain topic, it is likely to encounter others in the future as well.

The ONSIDE implementation from MobEmu also contains mechanisms for sorting the messages in a node's memory, in order to ensure that, when a node's memory is full and it has to download a new message, it will drop the least significant one that it has stored (instead of the oldest one, as per the default case). Furthermore, the

MobEmu implementation of ONSIDE also contains SENSE mechanisms for trust management and incentives.

4.4.11 Interest Spaces

Interest Spaces [53] is a context-adaptive and knowledge-based middleware for mobile collaborative systems. Its purpose is to offer a unified framework for data dissemination in ONs, providing mobile applications with a message exchange mechanism in networks where Wi-Fi access points or mobile broadband connections are not available, or an alternative is needed. MobEmu has implementations for its dissemination and reputation components.

A node's interest in the Interest Spaces framework is expressed as a tag, and there are three types of nodes: publishers, subscribers, and cache nodes. The former are nodes that can publish data. In order to do so, they must specify tags for the data objects they publish. Other nodes that are subscribed to those tags are able to receive messages generated by the publishers. Subscribers are nodes that are interested by information marked with certain tags. They are able to specify the tags they are interested in at any time, as well as to unsubscribe from them. Messages that subscribers are interested in should arrive as fast as possible, especially in environments where data can become stale quickly. Finally, the cache nodes represent the backbone of the Interest Spaces framework, in the sense that they perform the actual heavy lifting. Their task is to cache and transport data items for the benefit of others. Nodes are able to become cache nodes for certain tags when they are in the vicinity of other nodes interested in such tags, or when they are known to interact often with said nodes. They store data of interest to these nodes until they encounter them and deliver the data. A node is a cache node for a certain context, which is computed on the fly and can change very quickly.

The cache decision function is based on the idea that a node that sees a certain tag often (i.e., it encounters nodes that carry messages marked with that tag or that are interested in that tag) would be a good cache node for messages marked with that particular tag. Furthermore, this decision also takes into account the social connections between nodes. The trust and reputation component of Interest Spaces, entitled SAROS [54], is based on node gossiping for building trust in the other peers in the opportunistic network. Furthermore, messages that reach an interested node are not directly delivered to the application level. Instead, multiple copies of the same message (delivered on different routes by different nodes) are compared, and a quorum algorithm is used for selecting the valid version. This way, malicious nodes that alter messages can be detected, and the other nodes in the network are notified to avoid them.

4.5 Related Work

The most well-known simulator for opportunistic networks is ONE (Opportunistic Network Environment) [12]. It is a Java application that simulates various ON scenarios, allowing the customization of node behavior in terms of contacts, routing algorithms, transfer speed, battery, etc. The main functions of the ONE simulator are the modeling of node movement, inter-node contacts, routing, and message handling. Result collection and analysis are done through visualization, reports, and post-processing tools.

Each ONE node is represented by a main module which can connect to other submodules that represent its capabilities, such as radio interface, persistent storage, movement, energy consumption, or message routing. Transfer speed is abstracted to a communication range and bit-rate, which are statically configured and remain constant over the simulation. The node energy consumption model is based on an energy budget approach, where each node starts with a given energy level, which decreases when certain activities (such as data transmission) are performed.

Regarding node interactions, several approaches can be taken by a ONE simulation, similarly to MobEmu. Synthetic movement models can be employed, which include random movement models (such as Random Walk and Random Waypoint), map-constrained models (Random Map-Based Movement, Shortest Path Map-Based Movement, and Routed Map-Based Movement), as well as human behavior-based movement models (such as the Working Day Movement model presented in Sect. 2). Moreover, real-world traces such as the ones presented in Sect. 3 can be imported into the ONE simulator.

Messages are generated either randomly, or with a fixed source, destination, size, and interval. One main drawback of the ONE simulator is that all messages are unicast and directed, so they have a specific destination, which does not allow for data dissemination without further tweaking of the implementation.

There are six routing protocols included in the simulator, which can be run on a mobility model or a trace: direct delivery, first contact, Spray-and-Wait [55], PROPHET [56], MaxProp [57], and Epidemic [41]. However, new protocols can be added to ONE by extending a class (called *ActiveRouter*) and implementing specific methods.

The ONE simulator offers a Graphical User Interface (GUI) for visualizing the simulation live (showing the positions of nodes in the simulation space, their interactions, etc.), while statistics are also collected and exported in log files. Simulation scenarios can be built by defining the ON participants and their capabilities (such as storage capacity, transmission range and bit-rates, movement models, routing algorithms, etc.), as well as global parameters (such as duration, time granularity, etc.), through simple text-based configuration files. This offers less experienced users an easy way of creating and testing various scenarios.

The main caveat of ONE (and the reason we implemented MobEmu) is that it does not work for data dissemination out of the box. As stated before, messages can only be directed towards a single destination, so a publish/subscribe-based solution cannot be tested. Moreover, it does not offer support for community detection, social network knowledge, altruism modeling, or context data (i.e., topics of interest). Since we needed these components for implementing our routing and dissemination solutions, as well as our context-adaptive and knowledge-based middleware for mobile collaborative systems (Interest Spaces), we decided to implement our own ON simulator (MobEmu). We believe that not only does it offer more capabilities than the ONE simulator, but it is easier to maintain and extend, due to the simplicity and modularity of the code.

Other simulators that offer support for opportunistic networks include *ns-2*,[5] *ns-3*,[6] DTN2,[7] OMNet++,[8] or GloMoSim [58], but they have either been discontinued, are not used on a large scale, or do not offer complete support for ON simulations.

5 Conclusions

In this chapter, we have introduced the notions of mobility traces and synthetic models, highlighting the benefits and drawbacks for each of them. We argued that, when implementing an opportunistic routing or dissemination solution, it should be tested both using a social-based synthetic model, as well as on multiple mobility traces that cover as many real-life scenarios as possible.

In order to be able to do this, we presented MobEmu, an opportunistic network simulator that is able to run a mobility model or replay a trace, while applying the desired routing or dissemination algorithm. It is a Java application that offers simplicity and modularity, while at the same time allowing more experienced users a deep control over the scenarios they want to test. We showed that other similar simulators do not have all the capabilities that MobEmu offers, so its implementation was necessary.

Acknowledgements This chapter is based upon work from COST Action IC1406 High-Performance Modelling and Simulation for Big Data Applications (cHiPSet), supported by COST (European Cooperation in Science and Technology).

The research presented is also supported by national projects DataWay (PN-II-RU-TE-2014-4-2731) and MobiWay (PN-II-PT-PCCA-2013-4-0321).

References

1. Ciobanu, R.-I., Cristea, V., Dobre, C., Pop, F.: Big Data Platforms for the Internet of Things, pp. 3–34. Springer International Publishing, Cham, (2014)

[5] http://www.isi.edu/nsnam/ns/.

[6] https://www.nsnam.org.

[7] https://sites.google.com/site/dtnresgroup/home/code/dtn2documentation.

[8] https://omnetpp.org.

2. Pelusi, L., Passarella, A., Conti, M.: Opportunistic networking: data forwarding in disconnected mobile ad hoc networks. Comm. Mag. **44**(11), 134–141 (2006)
3. Gerl, Peter: Random walks on graphs. In: Heyer, Herbert (ed.) Probability Measures on Groups VIII. Lecture Notes in Mathematics, vol. 1210, pp. 285–303. Springer, Berlin Heidelberg (1986)
4. Johnson, D.B., Maltz, D.A.: Dynamic Source Routing in Ad Hoc Wireless Networks. In: Imielinski and Korth, (eds.) Mobile Computing, vol. 353. Kluwer Academic Publishers, (1996)
5. Nain, P., Towsley, D., Liu, B., Liu, Z.: Properties of random direction models. In: INFOCOM 2005. 24th Annual Joint Conference of the IEEE Computer and Communications Societies. Proceedings IEEE, vol. 3, pp. 1897–1907. March 2005
6. Haas, Z.J.: A new routing protocol for the reconfigurable wireless networks. In: Universal Personal Communications Record, 1997. Conference Record., 1997 IEEE 6th International Conference on, vol. 2, pp. 562–566. (Oct 1997)
7. Liang, B., Haas, Z.J.: Predictive distance-based mobility management for multidimensional pcs networks. IEEE/ACM Trans. Netw. **11**(5), 718–732 (2003)
8. Chiang, C.-C., Gerla, M.: On-demand multicast in mobile wireless networks. Proc. IEEE ICNP **98**, 14–16 (1998)
9. Camp, T., Boleng, J., Davies, V.: A survey of mobility models for ad hoc network research. Wireless Commun. Mobile Computing **2**(5), 483–502 (2002)
10. Derrida, B., Flyvbjerg, H.: The random map model: a disordered model with deterministic dynamics. J. de Phys. **48**(6), 971–978 (1987)
11. Soares, V.N.G.J., Rodrigues, J.J.P.C., Farahmand, F.: Impact Analysis of the Shortest Path Movement Model on Routing Strategies for VDTNs in a Rural Region. In: Proceedings of 7th Conference on Telecommunications, CONFTELE 2009
12. Keränen, A., Ott, J., Kärkkäinen, T.: The ONE Simulator for DTN Protocol Evaluation. In: Proceedings of the 2Nd International Conference on Simulation Tools and Techniques, Simutools '09, pp. 55:1–55:10, ICST, Brussels, Belgium, Belgium, 2009. ICST (Institute for Computer Sciences, Social-Informatics and Telecommunications Engineering)
13. Musolesi, M., Mascolo, C.: Designing mobility models based on social network theory. SIGMOBILE Mob. Comput. Commun. Rev. **11**(3), 59–70 (2007)
14. Boldrini, C., Passarella, A.: HCMM: modelling spatial and temporal properties of human mobility driven by users' social relationships. Comput. Commun. **33**(9), 1056–1074 (2010)
15. Watts, D.J.: Small Worlds: The Dynamics of Networks Between Order and Randomness. Princeton University Press, Princeton, NJ, USA (2003)
16. Ekman, F, Keränen, A., Karvo, J., Ott, J.: Working day movement model. In: *Proceedings of the 1st ACM SIGMOBILE Workshop on Mobility Models*, MobilityModels '08, 33–40, New York, NY, USA, 2008. ACM
17. Musolesi, M., Mascolo, C.: Mobility Models for Systems Evaluation. In: Garbinato, C., Miranda, H., Rodrigues, L. (eds.) Middleware for Network Eccentric and Mobile Applications, chap. 3, pp. 43–62. Springer Berlin Heidelberg, Berlin, Heidelberg, 2009
18. Barabasi, A.L.: The origin of bursts and heavy tails in human dynamics. Nature **435**(7039), 207–211 (2005)
19. Doci, A., Barolli, L., Xhafa, F.: Recent advances on the simulation models for Ad hoc networks: real traffic and mobility models. Scalable Computing: Practice and Experience, **10**(1), (2009)
20. Doci, A., Springer, W., Xhafa, F.: Impact of the Dynamic Membership in the Connectivity Graph of the Wireless Ad hoc Networks. Scalable Computing: Practice and Experience, **10**(1), 2009
21. Hummel, K.A., Hess, A.: Movement activity estimation for opportunistic networking based on urban mobility traces. In: Wireless Days (WD), 2010 IFIP, 1–5, Oct. 2010
22. Oliveira, J.G., Barabási, A.L.: Human dynamics: darwin and Einstein correspondence patterns. Nature **437**(7063), 1251 (2005)
23. Hidalgo R, C.A.: Conditions for the emergence of scaling in the inter-event time of uncorrelated and seasonal systems. Phys A: Stat. Mech. Its Appl. **369**(2), 877–883 (2006)
24. Vázquez, A.: Exact results for the Barabási model of human dynamics. Phys. Rev. Lett. **95**(24), 248701+ (2005)

25. Vázquez, A., Oliveira, J.G., Dezsö, Z., Goh, K.-I., Kondor, I., Barabási, A.-L.: Modeling bursts and heavy tails in human dynamics. Phys. Rev. E. **73**(3), 036127+ (2006)
26. Bigwood, G., Rehunathan, D., Bateman, M., Henderson, T., Bhatti. S.: Exploiting self-reported social networks for routing in ubiquitous computing environments. In: Proceedings of the 2008 IEEE International Conference on Wireless & Mobile Computing, Networking & Communication, WIMOB '08, pp. 484–489, Washington, DC, USA, 2008. IEEE Computer Society
27. Su, J., Scott, J., Hui, P., Crowcroft, J., De Lara, E., Diot, C., Goel, A., Lim, M.H., Upton, E.: Haggle: Seamless Networking for Mobile Applications. In: Proceedings of the 9th International Conference on Ubiquitous Computing, UbiComp '07, 391–408. Springer. Berlin, Heidelberg, 2007
28. Pitkänen, M., Kärkkäinen, T., Ott, J., Conti, M., Passarella, A., Giordano, S., Puccinelli, D., Legendre, F., Trifunovic, S., Hummel, K., May, M., Hegde, N., Spyropoulos, T.: SCAMPI: service platform for social aware mobile and pervasive computing. SIGCOMM Comput. Commun. Rev. **42**(4), 503–508 (2012)
29. Allen, S.M., Conti, M., Crowcroft, J., Dunbar, R., Lió, P.P.., Mendes, J.F., Molva, R., Passarella, A., Stavrakakis, I., Whitaker, R.M.: Social Networking for Pervasive Adaptation. In: Self-Adaptive and Self-Organizing Systems Workshops, 2008. SASOW 2008. Second IEEE International Conference on, pp. 49–54, (Oct. 2008)
30. Chaintreau, A., Hui, P.: Pocket Switched Networks: Real-world mobility and its consequences for opportunistic forwarding. Technical report, 2006 Computer Laboratory, University of Cambridge, February 2005
31. Chaintreau, A., Hui, P., Crowcroft, J., Diot, C., Gass, R., Scott, J.: Impact of human mobility on opportunistic forwarding algorithms. IEEE Trans. Mob. Comput. **6**(6), 606–620 (2007)
32. Hui, P., Crowcroft, J., Yoneki, E.: BUBBLE Rap: social-based forwarding in delay tolerant networks. In: Proceedings of the 9th ACM International Symposium on Mobile Ad Hoc Networking and Computing, MobiHoc '08, 241–250, New York, USA, 2008. ACM
33. Ciobanu, R.I.: Ciprian Dobre, and Valentin Cristea. Social aspects to support opportunistic networks in an academic environment. In: Proceedings of the 11th International Conference on Ad-hoc, Mobile, and Wireless Networks, ADHOC-NOW'12, 69–82. Springer, Berlin, Heidelberg (2012)
34. Hui, P., Chaintreau, A., Scott, J., Gass, R., Crowcroft, J., Diot, C.: Pocket switched networks and human mobility in conference environments. In: Proceedings of the 2005 ACM SIGCOMM Workshop on Delay-Tolerant Networking, WDTN '05, 244–251, New York, NY, USA, 2005. ACM
35. Hui, P., Yoneki, E., Chan, S.Y., Crowcroft, J.: Distributed community detection in delay tolerant networks. In: Proc. of 2nd ACM/IEEE inter. workshop on Mobility in the evolving internet architecture, MobiArch '07, 7:1–7:8, New York, NY, USA, 2007. ACM
36. Marin, R.-C., Dobre, C., Xhafa, F.: Exploring Predictability in Mobile Interaction. In: Emerging Intelligent Data and Web Technologies (EIDWT), 2012 Third International Conference on, 133–139. IEEE, 2012
37. Pietiläinen, A.-K., Oliver, E., LeBrun, J., Varghese, G., Diot, C: Mobiclique: Middleware for mobile social networking. In: Proceedings of the 2Nd ACM Workshop on Online Social Networks, WOSN '09, 49–54, New York, NY, USA, 2009. ACM
38. Srinivasan, V., Motani, M., Ooi, W.T.: Analysis and Implications of Student Contact Patterns Derived from Campus Schedules. In: Proceedings of the 12th Annual International Conference on Mobile Computing and Networking, MobiCom '06, 86–97, New York, NY, USA, 2006. ACM
39. Socievole, A., De Rango, F., Caputo, A.: Wireless contacts, Facebook friendships and interests: analysis of a multi-layer social network in an academic environment. In: Wireless Days (WD), 2014 IFIP, 1–7. IEEE, November 2014
40. Tsai, T.-C., Chan, H.-H.: NCCU Trace: social-network-aware mobility trace. Commun. Mag. IEEE **53**(10), 144–149 (2015)
41. Vahdat A., Becker, D.: Epidemic routing for partially-connected ad hoc networks. Technical report, Duke University, April 2000

42. Socievole, A., Yoneki, E., De Rango, F., Crowcroft, J.: Opportunistic message routing using multi-layer social networks. In: Proceedings of the 2Nd ACM Workshop on High Performance Mobile Opportunistic Systems, HP-MOSys '13, 39–46, New York, NY, USA, 2013. ACM
43. Moghadam A., Schulzrinne, H.: Interest-aware content distribution protocol for mobile disruption-tolerant networks. In: World of Wireless, Mobile and Multimedia Networks Workshops, 2009. WoWMoM 2009. IEEE International Symposium on a, 1–7, June 2009
44. Mtibaa A., Harras K.A.: Social-based trust in mobile opportunistic networks. In: 2011 Proceedings of 20th International Conference on Computer Communications and Networks (ICCCN), 1–6, July 2011
45. Ciobanu, R.I., Dobre, C., Toral, S.L., Johnson, P.: Jder: A history-based forwarding scheme for delay tolerant networks using jaccard distance and encountered ration. J. Netw. Comput. Appl. Daniel Gutierrez Reina **40**, 279–291 (2014)
46. Bigwood, G., Henderson, T.: Ironman: Using social networks to add incentives and reputation to opportunistic networks. In: SocialCom/PASSAT, pp. 65–72. IEEE, (2011)
47. Ciobanu, R.-I., Dobre, C., Dascălu, M., Trăuşan-Matu, Ş., Cristea, V.: SENSE: a collaborative selfish node detection and incentive mechanism for opportunistic networks. J. Network Computer Appl. **41**, 240–249 (2014)
48. Ciobanu, R.I., Dobre, C., Cristea, V.: Reducing congestion for routing algorithms in opportunistic networks with socially-aware node behavior prediction. In: Proceedings of the 2013 IEEE 27th International Conference on Advanced Information Networking and Applications, AINA '13, 554–561. IEEE Computer Society, Washington, DC, USA (2013)
49. Ciobanu, R.I. Dobre, C., Cristea, V: SPRINT: Social prediction-based opportunistic routing. In: 2013 IEEE 14th International Symposium and Workshops on a World of Wireless, Mobile and Multimedia Networks (WoWMoM), 1–7 June 2013
50. Ciobanu, R.I., Dobre, C., Cristea, V., Pop, F., Xhafa, F.: SPRINT-SELF: social-based routing and selfish node detection in opportunistic networks. Mobile Inf. Sys. **1–12**, 2015 (2015)
51. Ciobanu, R.-I., Marin, R.-C., Dobre, C., Cristea, V.: Interest-awareness in data dissemination for opportunistic networks. Ad Hoc Networks **25**(PB), 330–345 (2015)
52. Ciobanu, R.-I., Marin, R.-C., Dobre, C., Cristea, V., Mavromoustakis, C.X.: ONSIDE: socially-aware and interest-based dissemination in opportunistic networks. In: Network Operations and Management Symposium (NOMS), 2014 IEEE, 1–6 May 2014
53. Ciobanu, R.-I., Marin, R.-C., Dobre, C., Pop, F.: Interest spaces: a unified interest-based dissemination framework for opportunistic networks. J. Syst. Architec. **72**, 108–119 (2017)
54. Ciobanu, R.-I., Marin, R.-C., Dobre, C., Cristea, V.: Trust and reputation management for opportunistic dissemination. Pervasive Mob. Comput. **36**, 44–56 (2007). (Special Issue on Pervasive Social Computing)
55. Spyropoulos, T., Psounis, K., Raghavendra, C.S.: Spray and wait: an efficient routing scheme for intermittently connected mobile networks. In: Proceedings of the 2005 ACM SIGCOMM workshop on Delay-tolerant networking, WDTN '05, 252–259, New York, NY, USA, 2005. ACM
56. Lindgren, A., Doria, A., Schelén, O.: Probabilistic routing in intermittently connected networks. SIGMOBILE Mobile Computing Commun. Rev. **7**(3), 19–20 (2003)
57. Burgess, J., Gallagher, B., Jensen, D., Levine B.N.: Maxprop: Routing for vehicle-based disruption-tolerant networks. In: INFOCOM 2006. 25th IEEE International Conference on Computer Communications. Proceedings, 1–11, April 2006
58. Zeng, X., Bagrodia, R., Gerla, M.: GloMoSim: a library for parallel simulation of large-scale wireless networks. SIGSIM Simul. Dig. **28**(1), 154–161 (1998)

Virtualization Model for Processing of the Sensitive Mobile Data

Andrzej Wilczyński and Joanna Kołodziej

Abstract In this chapter, the k-anonymity algorithm is used for anonymization of sensitive data sending via network and analyzed by experts. Anonymization is a technique used to generalize sensitive data to block the possibility of assigning them to specific individuals or entities. In our proposed model, we have developed a layer that enables virtualization of sensitive data, ensuring that they are transmitted safely over the network and analyzed with respects the protection of personal data. Solution has been verified in real use case for transmission sports data to the experts who send the diagnosis as a response.

1 Introduction

1.1 Data Virtualization

Virtualization usually referrs to the situations where applications can use resources no matter where they are located, how they are stored or implemented and where they come from. Data virtualization is a variation of virtualization, where we can distinguish an abstract layer that provides a simpler interface and methods for accessing data. Data sources may be many, but the user who relies on this data will see one abstract layer. User does not have to know for instance what database languages are used to retrieve data from their physical storage, what type of API is used or what is the message structure. The end user may have the impression that this is one large database. Rick van der Lans describes data virtualization as follows: *Data*

A. Wilczyński (✉) · J. Kołodziej
Cracow University of Technology, Warszawska st 24, 31-155 Cracow, Poland
e-mail: and.wilczynski@gmail.com

A. Wilczyński
AGH University of Science and Technology,
al. Mickiewicza 30, 30-059 Cracow, Poland

J. Kołodziej
e-mail: jokolodziej@pk.edu.pl

J. Kołodziej et al. (eds.), *Modeling and Simulation in HPC and Cloud Systems*,
Studies in Big Data 36, https://doi.org/10.1007/978-3-319-73767-6_7

virtualization is the technology that offers data consumers a unified, abstracted, and encapsulated view for querying and manipulating data stored in a heterogeneous set of data stores, [1].

1.2 Mobile Data Virtualization

Still developing and increasingly sophisticated mobile applications need access to business data. The security of such data is very important and communication protocols must meet the appropriate trust level. There are two issues with mobile data transfer:

- Mobile Application Developers, which for mobile application developers is the standard for creating queries for downloading business data. This standard greatly improves application performance by omitting the integration process with each of the data providers separately.
- Development and Operations, due to the sensitivity of the data, provides the right level of security, easy access to data. Management is one of the most important issues.

Existing mechanisms for accessing business data from mobile applications are in most cases based on the API. However, there are some reputable faults associated with this type of solution. First and foremost, the API is different for each data provider and may change over time, so it is important to take care of the different ways of integration with these providers, which involves a large and continuous workflow. A typical mobile Api platform is an abstract backend-as-a-service (MBaaS) layer that defines a source and makes data available to potential mobile applications.

1.3 Mobile Cloud Computing Data Virtualization Security Issues

Nowadays, the use of mobile applications is practiced by the majority of the population. Applications are constantly expanding and require ever-increasing computing resources. Due to the fact that they are executed on mobile devices their performance has some limitations. This is related to limited energy and computing resources. It happens that the huge amount of data that they can deliver also does not fit on the device. According to the demand appeared the concept of mobile cloud, where data processing and storage can take place externally. This gives the ability to use very complex applications even on weak devices. It is therefore possible to build applications that deal with such operations/problems as image processing, natural language processing, crowd computing, GPS/Internet data sharing, sensor data applications, multimedia search, social networking, [2]. Mobile Cloud Computing (MCC) combines cloud computing with mobile cloud.

There are some security challenges that need to be addressed in this kind of systems, [3]:

- **Volume** - data is transferred between the different layers and then combined, which can cause problems in the integrity and inviolability of the data.
- **Velocity** - the speed of data collection forces the use of such encryption algorithms, which ensure the proper flow of data.
- **Variability** - data privacy must be ensured when data is no longer valid and should be deleted.

Each MCC system should ensure that the processing and transfer of data is consistent with the above security issues.

2 Motivation

Very often, data processing or computing operations on mobile devices consume large amounts of computing resources, which in turn puts a heavy burden on power and batteries. Processing of this data can take place in the cloud system and processing result can be returned to the mobile device as a response. It also happens that these data must be examined by experts so that further analysis is possible. These kind of data are often sensitive data that can not be shared with third parties. The main goal of this chapter is to design a model for sending sensitive data from mobile devices to cloud computing system (CC). Model should provide satisfying level of security and ability to analyze or process data by experts or third parties. They can not know the identity of the person who is associated with data due to compliance with data protection standards. The aim of this chapter is to design and implement a model meeting above requirements.

3 Related Work and Existing Solutions

One of the working examples is the approach proposed by Rocket Software. They provide integrated virtual views that allow direct access to mainframe data without having to transform them, called Rocket Data Virtualization (RDV). This technology enables real-time data usage without time–consuming ETL (extract, transform, load) operations. Typically, mainframes use log-based replication to locate operational data in System Management Facility (SMF). This type of data must be processed in such a way that it can be read in RDV, moreover, these data have to be transferred several times. The amount of data does not always allow you to quickly transfer them to the data warehouse. RDV delivers data in the right format without having to process it, so access to data is several times faster, [4].

Next approach of data virtualization is Red Hat JBoss Data Virtualization. This tool allows to combine data from heterogeneous environments, create virtual models and data access views, process them and share using simple interfaces. Data from multiple sources can be shared using SQL, such as JDBC or web services such as REST. For the user the data source is a one logical virtual model. Red HAT provides a graphical interface for easy creation this type of model. Each model allows to map source data to target formats required by end-applications. The application has the ability to integrate as data sources such tools as Hadoop, NoSql, SaSS, Data Warehouses and many type of files (xml, csv, excel), [5].

One possible solution for mobile data virtualization is the solution proposed by MadMobile. They define 3 elements that should comprise the mobile data virtualization platform:

- Mobile Data Sources
- Data Access APIs
- Mobile Data Catalog

The most interesting of the above three elements is Mobile Data Catalog, which is something like a repository that contains all the data sources. MadMobile has designed The KidoZen Mobile Data Virtualization Platform. This application creates a virtual representation for all data that can be downloaded by mobile applications, called Data Catalog. Each new element can be added to Data Catalog by specifying: data source, data source name, connector, operation, parameters, caching options. Next KidoZen makes the data available through the API using the Open Data Protocol, [6].

The next article shows the use of mobile cloud computing for real-time multimedia-assisted mobile food recognition application. The authors present there a mechanism for counting calories using CC. The main functions of the application are segmentation and image processing, and the use of deep learning algorithms to classify and recognize food. This type of operation due to the limitations of mobile devices can not be done directly on them. They use the Android capabilities for parting application activities into the front part installed on the mobile device and the backend where the application processing is done on the virtual Android image located in cloud, [7].

In the next chapter Mollah, see [8], describes the challenges and security issues in mobile cloud computing. He draws attention to aspects of MCC security, namely: cloud computing data security, virtualization security, partitioning offloading security, mobile cloud applications security, mobile device security, data privacy, location privacy and identity privacy. The general security requirements that are described by authors in this article apply to: confidentiality, integrity, availability, authentication and access control, privacy requirements.

4 Data Anonymization

Data anonymization (DA) is a process that allows data protection. In this process we can distinguish the mechanisms of encryption and deletion of personal data, which make it impossible to associate with individuals, provide their anonymity. Sensitive data transmitted over the network can be stolen and disclosed, which increases the risk of their use to the detriment of data providers. DA provides a high level of security even in the case of uncontrolled disclosure, because they can not be linked to the people they describe. EU has defined safety regulations for sensitive data. Given these regulations data can be divided into:

- Personal data - Data that allows direct identification of the person to whom they refer through the identification number or set of characteristics describing the person.
- Anonymous data - Data that can not be linked to the person to whom they relate. This can not be done by either the processor or any other person. After this process data is no longer personal data and is not subject to EU regulations on the protection of personal data.
- Pseudonymous data - after the process of anonymity there are some personal data, so they are still personal data. Nevertheless, the process of deanonimization is not an easy process, and it is difficult to identify the right person.

There are two approaches to DA:

1. K-anonymity - private tables contain a set of attributes that define and describe a person, if this set is available externally it is called a quasi-identifier.

Definition 1 *(k-anonymity requirement)* Each data sharing must be done so that each combination of quasi-identifier values can be matched to k different respondents.

Definition 2 *(k-anonymity)* Let $T(A_1, \ldots, A_m)$ be a table, and QI be a quasi-identifier associated with it. T is said to satisfy k-anonymity with respect to QI if each sequence of values in $T[QI]$ appears at least with k occurrences in $T[QI]$ [9].

To better illustrate the above definition, we show some example. Lets assume that we have the following table:

Table 1 consists 5 attributes and 8 records. There are 2 methods for obtaining anonymity for some k:

- Generalization - in this method, true attribute value is generalized, for instance the value of attribute "Date of birth" - 12071990 is replaced by 11061990 < Date of birth < 01011991.
- Suppression - in this method, some true values are replaced by asteriks'*'.

With respect to the quasi-identifiers $Dateofbirth$, Sex data has 2-anonymity, because we have at least 2 rows with the same attributes, with respect to the quasi-identifiers $Dateofbirth$, Sex, $Salary$ data has 1-anonymity, because we have single occurrences of values (Table 2).

Table 1 Private table

	Name	Date of birth	Sex	Profession	Salary
1	Melania Wolska	12071990	F	Teacher	2500
2	Kajetan Nowicki	21091987	M	Programmer	4000
3	Maja Nowak	21061957	F	Doctor	4000
4	Stanisaw Zakrzewski	25061946	M	Waitress	2000
5	Jan Zalewski	12071990	M	Waiter	2000
6	Igor Gorski	25061946	M	Plumber	2400
7	Zuzanna Nowakowska	21061957	F	Dentist	4400
8	Anna Baran	12071990	F	Doctor	4400
9	Wojciech Jakubowski	21091987	M	Teacher	4400

Table 2 Public table

	Name	Date of birth	Sex	Profession	Salary
1	*	12061990 < Date of birth < 12081990	F	*	2500
2	*	Date of birth > 21091986	M	*	4000
3	*	01061957 < Date of birth < 23061957	F	*	4000
4	*	Date of birth < 25061947	M	*	2000
5	*	12071989 < Date of birth < 12071991	M	*	2000
6	*	Date of birth < 25061947	M	*	2400
7	*	01061957 < Date of birth < 23061957	F	*	4400
8	*	12061990 < Date of birth < 12081990	F	*	4400
9	*	Date of birth > 21091986	M	*	4400

The algorithm described above has some drawbacks, it is susceptible to two types of attacks:

- Background Knowledge Attack - The case where the association of one or more quasi-identifier attributes containing sensitive values leads to a reduction in the range so that it can be deduced which record fits the individual.
- Homogeneity Attack - case where all sensitive values in set k records are identical.

2. l-Diversity - an extension of k-anonymity model, where the granularity of the data representation is reduced.

Definition 3 *(l-diversity)* The table is l-diverse if for every $q* - block$

$$-\sum_{s \in S} p_{(q*,s)} log(p_{(q*,s')}) \geq log(l) \tag{1}$$

where $p_{(q*,s)} = \frac{n_{(q*,s)}}{\sum_{s' \in S} n_{(q*,s')}}$ is the fraction of tuples in the $q* - block$ with sensitive attribute value equal to s [10].

Table 3 2-anonymous table

	Name	Date of birth	Sex	Profession	Salary
1	*	12061990 < Date of birth < 12081990	F	*	2500
8	*	12061990 < Date of birth < 12081990	F	*	4400
2	*	Date of birth > 21091986	M	*	4000
9	*	Date of birth > 21091986	M	*	4400
3	*	01061957 < Date of birth < 23061957	F	*	4000
7	*	01061957 < Date of birth < 23061957	F	*	4400
4	*	Date of birth < 25061947	M	*	2000
6	*	Date of birth < 25061947	M	*	2400
5	*	12071989 < Date of birth < 12071991	M	*	2000

Table 4 2-diversity table

	Name	Date of birth	Sex	Profession	Salary
1	*	01061957 ≤ Date of birth < 12081990	F	*	2500
8	*	01061957 ≤ Date of birth < 12081990	F	*	4400
2	*	Date of birth > 01061957	M	*	4000
9	*	Date of birth > 01061957	M	*	4400
3	*	01061957 ≤ Date of birth < 23061957	F	*	4000
7	*	01061957 ≤ Date of birth < 23061957	F	*	4400
4	*	Date of birth ≤ 01061957	M	*	2000
6	*	Date of birth ≤ 01061957	M	*	2400
5	*	01061957 ≤ Date of birth < 12071991	M	*	2000

Analyzing Table 3, you can see that the data are not susceptible to homogeneity attack. However, considering the records of 1 and 8, namely Melania and Anna. Melania may be Anna's neighbor whom she knows she was born on the same day as her, she also knows that she's earning 2500, so she can deduce that Anna earns 4400. After diversity process data are not susceptible to background knowledge attack as it presented in Table 4.

5 Data Exchange Model

We present in Fig. 1 the original contribution of this chapter is a model for transmission sensitive data via network. that model ensures that the receivers of data are not be able to assign them to real people, but they can send feedback to them. On the model we can distinguish the following elements:

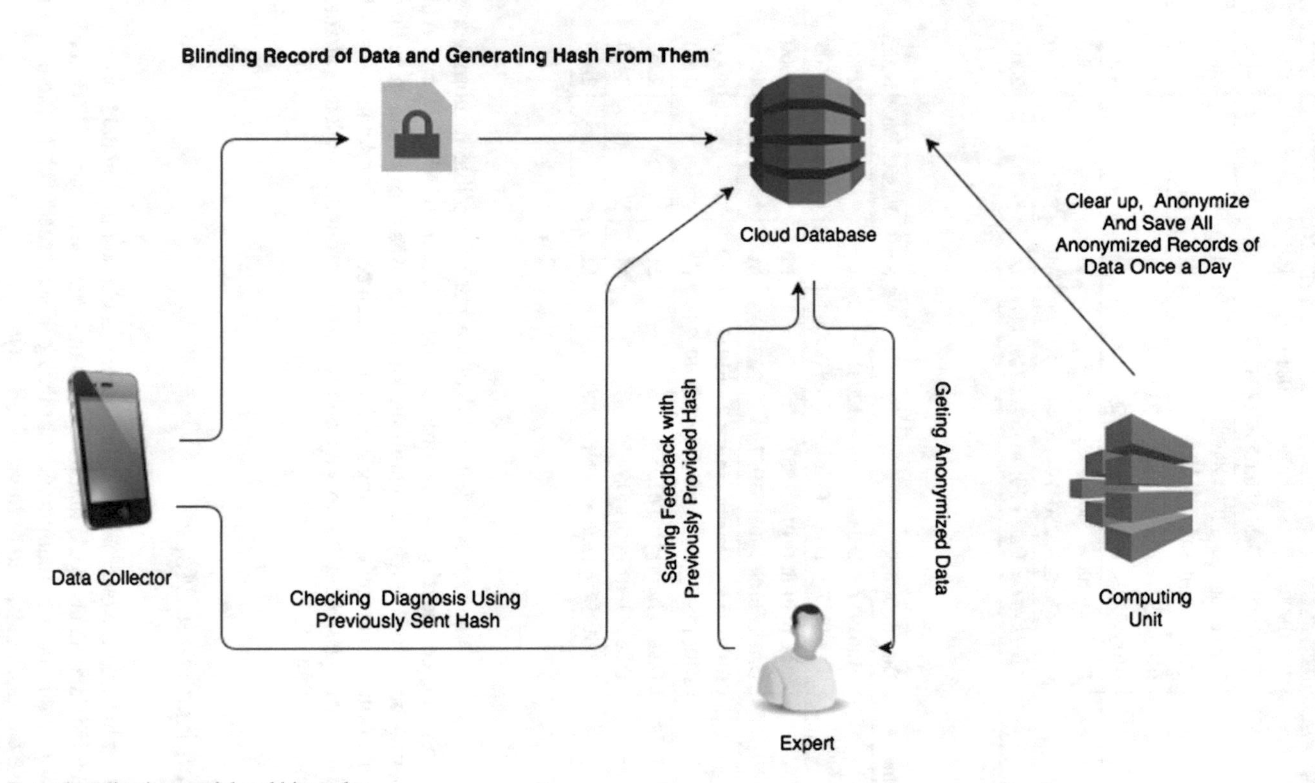

Fig. 1 Data virtualization model and hierarchy

- Data Collector - devices for collecting data from sensors;
- Cloud Database - database for storing blinded and anonymized data;
- Computing Unit - computational unit for decrypting and anonymizing data;
- Expert - person responsible for data analysis and feedback.

Data Collector, before sending data to the database, performs blinding operations on them using the pseudo one-time pad (OTP) algorithm. OTP is an encryption technique where message is paired with a random secret key [11]. Each character is encrypted by combining it with the adequate character from key. Algorithm uses a short key and as a result returning binary data string. In this work, algorithm has been called a pseudo OTP algorithm because the same key is used many times to encrypt and decrypt various messages, in the real OTP after encrypt and decrypt message the key is thrown and a new one is randomly generated. It is used only to blind sensitive data, which is sufficient at this stage. Encryption has been extended to binary data, which means that instead of characters, a binary key is used. After blinded operation with the MD5 function [12], the hash of the resulting data is generated.

Blinded data together with hash goes to the Cloud Database. Once in a while, all data records from database are downloaded by the Computing Unit and cleared up using the same key that was used to blind them. After that, data are anonymized and stored in the database, allowing them to be analyzed. Expert analyzes sensitive data and sends the feedback, which is stored under the same hash as the blinded data. Expert does not know who the data belongs to, which ensures compliance with the personal protection laws.

After a certain period of time, data collector performs a database query to verify the result of the analysis of the previously transmitted data. The same hash that was previously generated from the blinded data is used for the search. If the result has been already saved by an expert, data collector gets it, otherwise the query is repeated in a while.

6 Model Use Case and Implementation

To illustrate the presented model, we prepared an implementation of the system for the collection and analysis of sports data transmitted by athlete's sensors. These sensors collect information about the heart rate of the person performing the exercise and send them to the CC. As a result, the person receives a diagnosis of the time spent in each exercise zone during the training. The data used for the analysis is the real data collected from training of 19 people. Implementation was done in JAVA.

It can be observed from Table 5 that we have 8 attributes describing each person. The heart rate attribute was cut off due to the very long representation.

Table 6 presents blinded data together with generated MD5 hash. Each attribute is first saved in JSON format, then string of JSON format is blinded. Below we present fragment of json for a single record:

Table 5 Sport data

	Name	Sport	Weight	Height	Heart rate (resting)	Age	Duration (min)	Heart rate
…	…	Running	…	…	…	…	…	…
8	Barreta Page	Running	51	156	90	39	115	0, 90, 92, 98, 97, 97, 99, 106, 112, 116…
9	Lens Temple	Cycling	74	162	91	37	115	0, 0, 170, 151, 0, 196, 196, 196, 196…
10	Bevise Kenyon	Cycling	81	167	82	35	158	0, 221, 87, 91, 102, 109, 107, 103…
11	Adolphe Nigellus	Cycling	88	174	83	35	174	0, 111, 110, 112, 115, 116, 116, 115…
12	Bentleye Nelson	Cycling	55	178	85	34	115	0, 0, 0, 75, 206, 199, 201, 201, 154…
…	…	Running	…	…	…	…	…	…

Table 6 Blinded sport data

	Hash	Data
	…	…
8	782d74c56dbe99d6cc494a1c7284305d	00000010 0011110 00011101 00001100 00001011 000110…
9	eaf29fdefc33c8780f53e1de61c88d52	00000010 0011110 00011101 00001100 00001011 000110…
10	4fb95b92fa76bdf9cd208d05dafb90b4	00000010 0011110 00011101 00001100 00001011 000110...
11	1e11b033a1239321a4898dd1320d5428	00000010 0011110 00011101 00001100 00001011 000110...
12	b1580545d102ca661c4dfacf292a9278	00000010 0011110 00011101 00001100 00001011 000110...
	...	...

{"*duration*" : 115, "*heartrate*" : "00, 0, 170, 151, 0, 196, 196, 196, 196, 196, 196…", ...}

In implementation we used ARX library for anonymization [13]. In this library we can distinguish 4 types of attributes: insensitive, sensitive, quasi-identifying and identifying. Our use case classifies the attributes as follows:

- {sport, heart rate (resting), duration, heart rate} - insensitive
- {name} - identifying
- {weight, height, age} - quasi-identifying

For weight, height and age generalization has been used. For each quasi-identifier attribute we have prepared individual generalization hierarchy using intervals. As a

Table 7 2-anonymous sport data

	Name	Sport	Weight	Height	Heart rate (resting)	Age	Duration (min)	Heart rate
...	...	...	...	...	...	...	...	...
8	*	Running	[48, 60[	*	90	[36, 45[	115	0, 90, 92, 98, 97, 97, 99, 106, 112, 116...
9	*	Cycling	[72, 84[	*	91	[36, 45[	115	0, 0, 170, 151, 0, 196, 196, 196, 196...
10	*	Cycling	[72, 84[	*	82	[24, 36[	158	0, 221, 87, 91, 102, 109, 107, 103...
11	*	Cycling	[84, 96[	*	83	[24, 36[	174	0, 111, 110, 112, 115, 116, 116, 115...
12	*	Cycling	[48, 60[	*	85	[24, 36[	115	0, 0, 0, 75, 206, 199, 201, 201, 154...
...	...	...	...	...	...	...	...	...

model 2-Anonymity has been adopted. The weights of the individual attributes are respectively: 0.5, 0.5, 0.5. Data presented in Table 7 have been anonymized in such a way that the expert has still no problems with a diagnosis.

7 Simulation Analysis

The implementation has shown that it is possible to send sensitive data meeting appropriate standards without losing their quality and consistency. These data can be sent to the cloud computing where their processing or analysis can be carried out independently. This approach provides the right level of security and enables performing computational operations or processing large amounts of data type of Big Data in an independent cloud computing environment. As we can see on Table 8 the whole process of assuming blinding and anonymization of data is done very fast. Simulation was performed on a computer MacBook Pro, 2,7 GHz Intel Core i5, 8 GB 1867 MHz DDR3.

Amount of data is small because in implementation we did not want to use data generator, the real data was used. Access to this type of data is very limited. However, the results we received are satisfactory and well predicted for the future.

Table 8 Process execution times

Operation type	Average time (miliseconds)
Blinding one record of data	744
Anonymization 19 records of data	320

8 Conclusions and Future Work

Data anonymization can support processes where sensitive data is sent. In the proposed model we have provided the ability to analyze data through external resources without disclosing to whom those data belong. Experts or analysts do not need to know the identities of the people they diagnose, all they have to do is to send feedback to them. Model can also be implemented in different environment. For instance, in the cloud computing in task scheduling, where the scheduler does not need to know who orders task and for what task has to be executed.

The future work will focus on optimizing the selection of hierarchies for generalizing attributes and their weights. We are considering the use of Stackelberg game [14] to decide how best to match these parameters. Currently, this process is done manually and should be automated.

Acknowledgements This chapter is based upon work from COST Action IC1406 High-Performance Modelling and Simulation for Big Data Applications (cHiPSet), supported by COST (European Cooperation in Science and Technology).

References

1. van der Lans, R.: Data Virtualization for Business Intelligence Systems: Revolutionizing Data Integration for Data Warehouses. Morgan Kaufmann Publishers Inc., San Francisco (2012)
2. Fernando, N., Loke, S.W., Rahayu, W.: Mobile cloud computing: a survey. Future Gener. Comput. Syst. **29**, 84106 (2013)
3. Jakóbik, A.: Big Data Security. Springer, Berlin (2016). https://doi.org/10.1007/978-3-319-44881-7_12
4. Software, R.: Rocket data virtualization. PDF document, http://www.rocketsoftware.com/sites/default/files/resource_files/DS_Data_DVS%20012615.pdf?flag=meta&product=rocket-data-virtualization&family=rocket-data&solution=data-virtualization&resourcetype=datasheet&resourcebn=rocket-data-virtualization&resourcefbn=DS_Data_DVS%20012615.pdf
5. Redhat: Jboss data virtualization. Electronic document, https://developers.openshift.com/jboss-xpaas/data-virtualization.html
6. Kidozen: From mdm to mdm. PDF document, http://www.kidozen.website/wp-content/uploads/2015/12/Mobile_Data_Virtualization.pdf
7. Pouladzadeh, P., Peddi, S.V.B., Kuhad, P., Yassine, A., Shirmohammadi, S.: A virtualization mechanism for real-time multimedia-assisted mobile food recognition application in cloud computing. Cluster Comput. **18**, 10991110 (2015)
8. Mollah, M., Azad, M.A.K., Vasilakos, A.: Security and privacy challenges in mobile cloud computing: Survey and way ahead. J. Netw. Comput. Appl. **84**, 3854 (2017)
9. Ciriani, V., De Capitani, S., di Vimercati, S., Foresti, P.Samarati: k-anonymity. secure data management in decentralized systems. Adv. Inf. Secur. **33**, 323353 (2007)
10. A. Machanavajjhala J. Gehrke, D.Kifer, M. Venkitasubramaniam: l-diversity: Privacy beyond k-anonymity. Proceedings of the 22nd International Conference on Data Engineering, 2006. ICDE '06. pp. 24–24 (2006)
11. Bellovin, S.M.: Frank miller: Inventor of the one-time pad. Cryptologia **35**(3), 203–222 (2011). https://doi.org/10.1080/01611194.2011.583711
12. Preneel, B.: Cryptographic Hash Functions: Theory and Practice. Springer, Berlin (2010). https://doi.org/10.1007/978-3-642-17401-8_9

13. Prasser, F., Kohlmayer, F.: Putting Statistical Disclosure Control Into Practice: The ARX Data Anonymization Tool. Springer, Berlin (2015)
14. Jakóbik, A., Wilczynski, A.: Using polymatrix extensive stackelberg games in security aware resource allocation and task scheduling in computational clouds. J. Telecommun. Inf. Technol. **1**, 71–80 (2017)

Analysis of Selected Cryptographic Services for Processing Batch Tasks in Cloud Computing Systems

Agnieszka Jakóbik and Jacek Tchórzewski

Abstract This chapter evaluates the features and a computational load of two proposed cryptographic procedures which aim to protect confidentiality and data integrity in Cloud Computing (CC) systems. It should be kept in mind that a bad use of some cryptographic tools may negatively impact the overall CC operation. Regarding this, meeting the Quality of Service (QoS) requirements is only possible when the security layer applied does not interrupt the computing process. The security layer applied to tasks should also fulfill the advanced security conditions present in CC systems. Thus, the solutions aiming to protect both the user data as well as the whole system have to deliver the scalability, multi-tenancy and complexity that these systems demand. We present a cryptographic service based on blind RSA algorithm and Shamir secret sharing that supports batch tasks processing. Hence, this service is suitable for CC systems equipped with a monolithic central scheduler and many Virtual Machines (VMs) as working nodes. Blind RSA cryptographic system is used to encrypt the data without actually knowing any details about the tasks content. Shamir secret sharing procedure is proposed in order to assure whether all VMs in the system gave back their shares after deploying the batch of tasks on them or not.

1 Introduction

Cloud Computing environments are very intensively used by private, academic and commercial organizations. They may offer combined solutions where many wide-range services and systems are offered, [1–3] as well as dedicated solutions which focus on certain problems [4].

There are a lot of security domains in CC developed so far [5]. Assuring the proper security level of the infrastructure, applications, user access, data and provider is

A. Jakóbik (✉) · J. Tchórzewski
Tadeusz Kościuszko Cracow University of Technology, Warszawska 24, Cracow, Poland
e-mail: akrok@pk.edu.pl

J. Tchórzewski
AGH University of Science and Technology Krakow, Mickiewicza 30, Cracow, Poland
e-mail: jacek.tchorzewski@onet.pl

J. Kołodziej et al. (eds.), *Modeling and Simulation in HPC and Cloud Systems*,
Studies in Big Data 36, https://doi.org/10.1007/978-3-319-73767-6_8

essential for building any trustworthy system. Additionally, such responsibilities are also enforced by international regulations and norms. For instance, by The Data Protection Directive, number 95/46/EC, [6], which regulates the processing of personal data in the European Union. As in any IT system, CC systems have to assure privacy, confidentiality, integrity and availability of the data and services involved. All traditional cryptography tools, such as: symmetric-key cryptography, public-key cryptography, and crypto systems may be used. Among them: RSA encryption, Schnorr signature, El-Gamal encryption, PGP standard, electronic cash systems, signcryption systems, and systems for secret sharing [7].

Numerous additional features of the CC systems influence the whole cryptographic process. The following features may be defined as the most important for the system reliability:

- **Elastic re-provisioning**. The addition or expansion of resources result in a dynamic behavior in terms of the number of users, services, etc.
- **Virtualization**. The virtualization of physical resources requires dedicated techniques in order to assure that the security constraints are met.
- **Competitiveness**. The optimization of the cryptographic algorithms is required to reduce the costs, which leads to achieve business profitability.
- **Delocalization**. A service used in many different geographical points may require considering various user profiles with different security levels attached to them.
- **Multitenancy**. Additional security layers may be necessary when multiple users can work on the same data simultaneously due to resource sharing [8].

It may be observed that any cryptography protocol has to be examined as far as its features and scalability are concerned. In addition, it has to be designed to act accordingly to two main scenarios: Data in transit, and Data at rest. This behavioral differentiation lowers the total costs of task processing and storage. Properly optimized cryptographic algorithms should also take into account the features of the virtual resources that will be used during the data processing. Additional constraints may be incorporated for Big-Data scenarios [9], and when sensitive data is involved [10].

In this work, the authors examined the cryptographic solution based on the non-conventional usage of two well known algorithms, thus, the Rivest - Shamir - Adleman cryptosystem (RSA), and the Shamir's secret sharing algorithms. The RSA public-key encryption scheme algorithm was used with a previous blinding step, instead of just encode the raw data. The Shamir secret sharing was proposed in order to assure the data completeness, instead of a traditional hash function. Such a service allows to check data confidentiality and integrity by the means of lightweight cryptographic algorithms. The system was designed to preserve the anonymity of the obtained results outside the working nodes. Moreover it allows to encrypt the data assuming not trustworthy data storage units. Hence, data storage units are responsible for encoding all non-plain-text data.

On the other hand, intensive cryptographic operations are computed outside the main task working nodes. It allows these nodes to be ready for arriving jobs as soon as possible.

In this chapter, we also presents an extensive scalability analysis of the proposed solution. All the experiments were performed on CloudSim simulator. Due to a very complex nature of computer networks and distributed systems, simulation-based approaches to the performance analysis of such systems have been widely applied, [11–13]. CloudSim tool was configured in order to simulate the features of Amazon Cloud VMs.

Major contributions of this chapter include:

1. Description of the proposed cryptographic solution, adapted to CC systems.
2. Implementation and experimentation of the proposed algorithms, scheduling process and tasks.
3. Extensive performance analysis of the proposed cryptographic service on the task-processing flow.

The chapter is structured as follows. Firstly, the authors describe the batch task execution process in Cloud Computing systems in Sect. 2. Then, in Sect. 3, a security requirements mapping and a task security layer are described. In Sect. 4, the proposed cryptography service is explained in detail. Experimental results performed in order to evaluate the proposed model are presented in Sect. 5. In Sect. 6, the authors summarize the chapter and propose key topics for further research.

2 Batch Task processing in Cloud Computing systems

The Cloud Computing paradigm allows sharing services and resourses by offering a wide variety of services at different levels, such as: virtual disks, image libraries, data storage systems, scientific computing services, Internet of Things solutions, High Performance Computing, Backup and Recovery systems, data archiving disks, Big Data processing frameworks, and security services [8].

They offer services and resources that may be characterized by:

- virtualization - providers are delivering a virtual (not physical) computer hardware platform like virtual machines, storage devices in the form of virtual disks, and virtual computer network resources;
- multitenancy - in CC model resources are shared among many users,
- massive scalability on demand - the ability to change for example computing capacity, bandwidth and storage space, in the time that may be chosen be the user;
- flexibility - users can chose the Cloud services and solution on demand,
- pay as you go - the payment for for using services depends on the time they consumed,
- self-provisioning of resources - additional elements i.e. Virtual Machines, storage disks and other resources may be added to the pool of uses resources any time.

The concept of task, understood as the unit that represents the work delivered to the CC system to be processed by its working nodes, enables to express the general concept of jobs and services offered by CC. Unlike long running processes, which are run until an operator or an automatic supervisor kill them, aforementioned tasks run for a determined amount of time in order to process their given instructions and then finish. Various types of tasks are run using their own containers linked to users resources. Among them:

- Performing a data-processing script.
- Backing-up data base.
- Updating NoSQL database service.
- Compiling source code, running tests, and producing software packages.
- Performing complex data flows.
- Running Distributed Denial of Service (DDoS) protection service.

Every task is identified by a unique name or ID and is executed with parameters specified in the task definition. Tasks completion may may be dependent on other tasks, which can be referenced by their name or ID, or on additional data. The task definition may include access management rules to govern the access to required resources, such as: the memory and CPU, task input file size, and task output file size. In addition, it also specifies several properties of the container that may host the task, such as the CPU architecture and GPU details, as well as other environmental variables, such as security requirements.

Regarding tasks scheduling and processing, theses tasks may be processed as single jobs or in a batch way. On one hand, processing each task as a single job means that this task is computed independently of the other tasks. On the other hard, processing tasks in a batch way means that the tasks that are packed together in a batch are to be computed as parallel jobs. In this case, in order to further process the batch, all tasks in the batch have to be successfully computed and finished. Hence, the first tasks completed have to wait for the last tasks to finish. Then and only then the batch may be post-processed, delivered to the end-user or to another service. Map-Reduce tasks may be a good example of batch tasks.

In order to deploy the submitted tasks on the resources while meeting tasks' requirements and following a set of rules and heuristics, a resource scheduler is usually used, such as the default AWS Batch Scheduler, which is based on FIFO queues. Cloud providers allow the usage of several scheduling approaches. Among them: Automated scheduling, Manual scheduling, and Custom scheduling.

Like all the large-scale systems, a CC system may be architecturally described as the following set of modules:

- **Task Gathering Unit (TGU)**. This module collects the tasks from CC end-users and packs them in batches.
- **Dealer unit (SD)**. This module sends batch of tasks to the workers through its scheduler (S), which perform the spreading logic between working nodes, according to the chosen objective and tasks requirements.
- **Worker nodes (SW)**. This module is responsible for actually perform task computation, and usually is composed of a set of VMs.
- **Storage Center (SC)**. This module stores the computing results for further processing or to serve them to end-users, as described in Fig. 1.

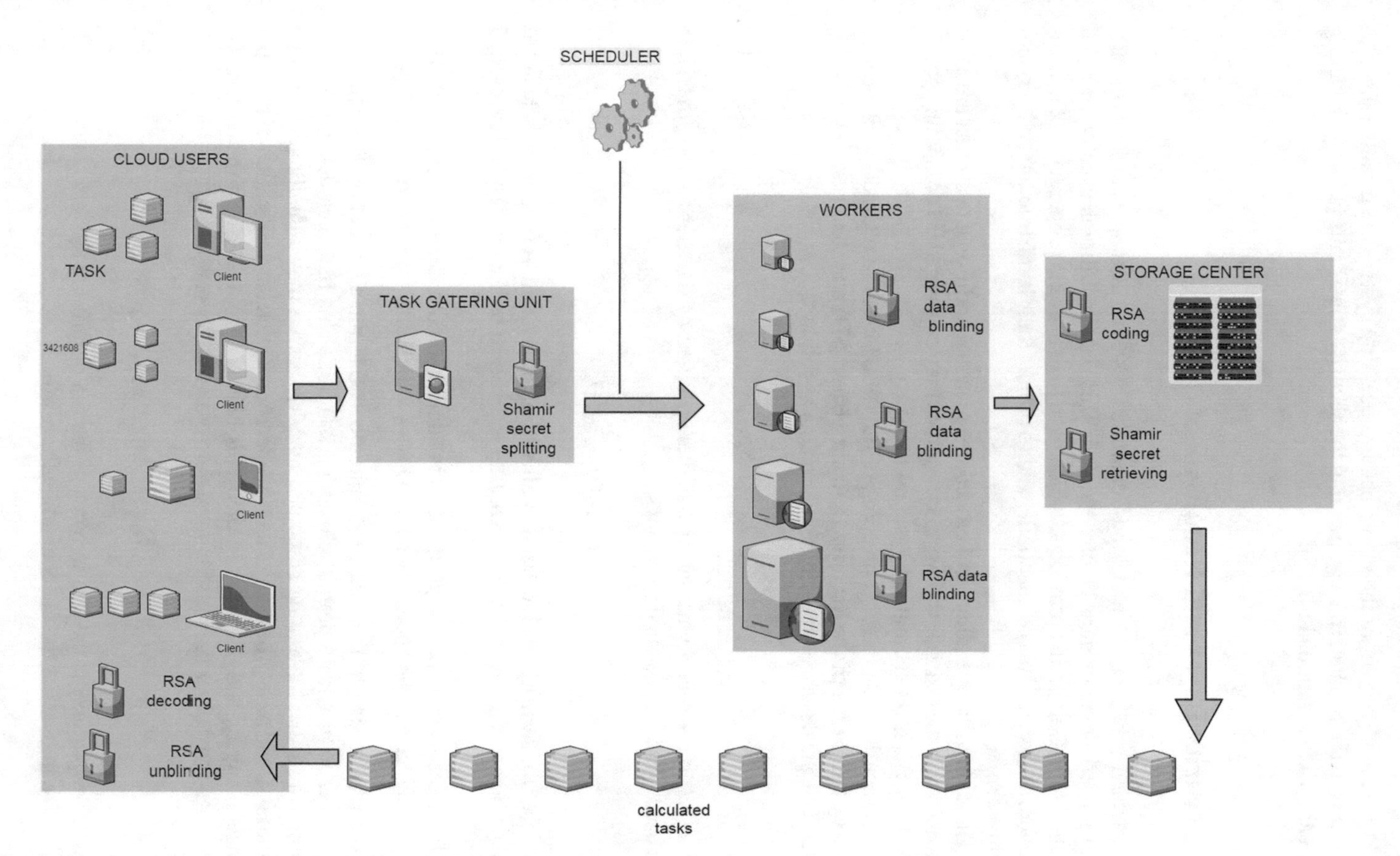

Fig. 1 The proposed cryptographic service for CC architecture

The main objectives of the whole batch processing system are usually related to minimizing the batch makespan, meeting the security constraints [14], and lowering the energy consumption [14, 15].

3 Security Layer and Requirements

Guaranteeing the security in a Cloud Computing system is a complex task since many inter-related elements must be orchestrated. This responsibility is usually managed by the Cloud Computing providers, which have to deliver the proper level of authentication, authorization, confidentiality and to protect their infrastructure and services from external attacks.

These security operations and procedures generate additional costs that are to be considered in the task scheduling process, since they consume VM computing power as well as the tasks submitted by end users.

The security requirements mapping of the Cloud users into Cloud VMs may be done by the means of the Security Demand Vector (SD) and the Trust Level Vector (TL). The SD describes security requirements as follows:

$$SD = [sd_1, \dots sd_n] \tag{1}$$

where sd_j is the security demand parameter specified by the user for the jth task in the batch. The TL, denoted as follows:

$$TL = [tl_1, \dots, tl_m] \tag{2}$$

describes the security levels offered by all VMs in the system. A task can be scheduled to a particular VM when it offers a security level equal or higher than that demanded by SD.

Each security service or protocol adds additional computational overhead. This security layer may be applied in a biased way, which may be divided into two parts:

- The bias related to the security requirements that must be met by the scheduler in order to provide a proper scheduling solution. This bias includes the security operations that have to be computed by VMs before they actually start running the task, and post-processing security operations, such as: verification of the input data integrity, and encoding results. This bias is described as estimated time (s) required to deliver the security level demanded and is denoted by:

$$b(sd_j, wl_j, tl_i, cc_i, inputSize_i, outputSize_i) \tag{3}$$

where $inputSize_i$ is the size in bytes of the file describing the task, $outputSize_i$ is the size in bytes of the file containing the result.

The following Security Bias Matrix (SB) is therefore obtained by representing aforementioned biases as a matrix:

$$SB(SD, TL) = [b(sd_j, wl_j, tl_i, cc_i, inputSize_i, outputSize_i)]_{i=1,2...,m}^{j=1,2...,n} \quad (4)$$

- The rest of security processes, protocols and operations that have an impact on the batch makespan, such as: security protocols performed by the Task Gathering Unit in order to send the tasks to the Worker Nodes, cyphering the data stored in the data center, and verification of the digital signature of the end user who wants to recover some results from the Cloud Computing system. Let us denote this kind of bias by B.

4 Proposed Cryptography Service

4.1 *Blind RSA Tasks Storage Service*

RSA (Ron Rivest, Adi Shamir and Leonard Adleman) is a popular public key algorithm for encrypting and decrypting data [7]. It is well known that data encryption allows to store the data at rest in a secure way. However, some data, such as task results, may need to be sent through the data center, since this data is not always stored on the working nodes which run the tasks but stored in dedicated virtual disks or data warehouses.

In order to solve both problems, thus, data in motion and data at rest, the blinded version of the RSA algorithm is used. The data in motion that is sent from SW to SC is not ciphered but only blinded in order to protect it from unauthorized access, what is a much more light-weight procedure. The data at rest in SC are hardly encoded.

Therefore, the SC is not longer aware of the exact data that will be encoded and stored due to the blinding process, and therefore it works as a black-box coding engine. Regarding an authorized final user, retrieved results are to be decoded when retrieved from the SC.

The blind RSA algorithm is executed in three steps:

1. *Key preparation*. Firstly, two different large random prime numbers p and q must be chosen by the user. Then, $n = pq$, the modulus for the public and private keys, is computed. The next stage needs: $\phi(n) = (p - 1)(q - 1)$ to be computed. Then, an integer e such that $1 < e < \phi(n)$ and e is co-prime to e have to be found. It means that e, and $\phi(n)$ share no factors other than 1, that is:

$$gcd(e, \phi(n)) = 1. \quad (5)$$

The public key, which is sent to the SP, have two elements: the modulus n and the public exponent e. The private key is composed of the modulus n and the private exponent d:

$$d = e^{-1}(mod(n)). \tag{6}$$

The private key is kept by the user and should never be revealed. The public key may be stored in a public data base. It is sent through the Cloud Computing system to the unit that will use it.

2. *Message blinding*. In order to blind the data $m \in \{0, 1, \ldots, n\}$, m has to be multiplied by $k^e modn$, where the blinding factor, denoted as k, is randomly chosen. Then the SW sends the blinded message $m * k^d modn$ to the SC and the blinding coefficient to the user.
3. *Message coding*. Next, the SC has to cipher and store the blinded message: $c = (m * k^d)^e modn$.
4. *Message unbliding and uncoding*. Finally, the message may be unblinded by the user:

$$c * k^{-1} modn = (m * k^d)^e * k^{-1} modn = \tag{7}$$

$$m^e * k^{de} * k^{-1}(modn) =$$

$$m^e * kk^{-1}(modn)$$

and deciphered:

$$(m^e)^d (modn) = m$$

4.2 Batch Completeness Verification Based on Secret Sharing

Tasks in the same batch are computed and stored in different locations. Nevertheless, to further process the batch or to send results to the user, the CC system have to guarantee that all the elements of the batch were collected. The proposed scheme may be the alternative for traditional methods regarding information integrity checking based on hash functions, such as SHA-2 [16]. Secret sharing procedures allow to spread certain knowledge among a group of participants, so that only a given set of them are able to recreate the knowledge by collecting and combining their shares [17]. Let us assume a batch of tasks B that is run by a set of worker nodes (SWs) t, being $t < n$, where n denotes the whole cluster. In order to check whether the result of this batch B is complete or not, the secret must be recreated from the set of machines t. It is also guaranteed that a subset s, being $s < t$, can not recreate this secret. Let be D the first unit participant of the secret and n the number of participants in the system. Let us assume the batch of tasks B was deployed on $t - 1$ VMs. Hence, each VM in that set has at least one task from the batch to be run. The Shamir scheme uses polynomial interpolation over finite field $GF(q)$, where $q >= n + 1$.

1. *Secret splitting*. The dealer choses n distinct nonzero elements from $GF(q)$:

$$x_1, x_2, \ldots, x_n$$

   and allocates them among participants, and sets the element $K \in GF(q)$ as the secret. The shares of the secret are therefore created by based on the scheme:

   - The dealer sets the elements

   $$a_1, a_2, \ldots, a_{t-1} \in GF(q)$$

     randomly, uniformly and independently.
   - if the equation

   $$a(x) = K + a_1 x + a_2 x^2 + \cdots + a_{t-1} x^{t-1} \tag{8}$$

     is polynomial of degree $t - 1$ then the shares are defined as $y_i = a(x_i)$ for i $= 1, 2, \ldots n$.

2. *Retrieving the secret.* If all $t - 1$ participants gather their shares together, they formulate the set of $t - 1$ points (x_i, y_i) of the polynomial a. Using Lagrangian interpolation [18], it is possible to find the unique polynomial of degree $t - 1$ passing throughout that points. The secret may be therefore found by taking the value of that polynomial at point 0. The shares may be also computed from the system of linear equations, for example using the Gauss Elimination method [19]:

$$y_1 = K + a_1 x_1 + a_2 x_1^2 + \cdots + a_{t-1} x_1^{t-1} \tag{9}$$

$$y_2 = K + a_1 x_2 + a_2 x_2^2 + \cdots + a_{t-1} x_2^{t-1} \tag{10}$$

   ...

$$y_{t-1} = K + a_1 x_t + a_2 x_t^2 + \cdots + a_{t-1} x_t^{t-1} \tag{11}$$

   where y_i for i $= 1, 2, \ldots$ i $- 1$ are the known shares,

$$K, a_1, a_2, \ldots, a_{t-1} \in GF(q)$$

   are the unknown. The solution of this system, including the secret K may be found only if exactly $t - 1$ participants gave their share. In the proposed service we assumed $t - 1 = n$, that means that all SW are sharing the secret.

The steps of the combined algorithm are the following, as shown in Fig. 1.

1. The unit D prepares the Shamir shares for all SW,
2. The shares are then given to the SW, as well as the schedule to be executed,

3. The SW computes the tasks and blinds the result by the means of the RSA blinding scheme. The blinded results are then sent to the SC, while the unblinding number is sent to the user.
4. The SC retrieves the Shamir secret in order to verify if all SWs gave their results. If so, the SC codes the blinded result using RSA cipher and stores it.
5. The SC sends the encoded result to the end user under his request, who can decode and unblind it.

5 CloudSim Experimental Results

5.1 Test Bed

The tests were run using CloudSim framework for modeling and simulation of CC infrastructures and service [20, 21].

The cluster was simulated by extending the data center entity present in CloudSim. Due to this, all VMs in the experiment are deployed on a sible physical host. The data center broker works with the non deterministic central scheduler described in [24, 25]. The data gathered from the experiment was collected by using a multi-agent monitoring system [26], and the function used to assign tasks to VMs was *bindCloudletToVm*. CloudSim worker nodes were designed by modeling the features present in M3 EC2 Amazon instances [27] as shown in Table 1.

Table 1 Mapping M3 EC2 Amazon instances into testing bed, *mips* denotes computing capacity, *ram* denotes VM memory in MB, CloudSim PE (Processing Element) class represents CPU unit, defined in terms of Millions Instructions Per Second (MIPS) rating, *pesNumber* represent the number of Pe elements in each VM Equivalent CPU speed is the clock speed of Intel Core i7 64-bit x86-64 processors with equivalent single thread performance

Amazon t				
Instance	vCPU	Mem (GB), [22]	SSD Storage (GB)	Eq. CPU speed, [23]
m1.small	1	1.6	160 GB HDD	0.71 GHz
m1.medium	1	3.75	410 GB HDD	1.43 GHz
m3.large	2	7.3	32 GB SSD	2.48 GHz s
m3.2 × large	8	30	2 * 80 GB SSD	2.38 GHz
CloudSim				
VM type	*pesNumber*	*ram* (MB)	*size* (MB)	$cc = mips$ (MIPS)
m1.small	1	1600	163840	710
m1.medium	1	3750	419840	1430
m3.large	2	7300	32768	2480
m3.2xlarge	8	30000	163840	2380

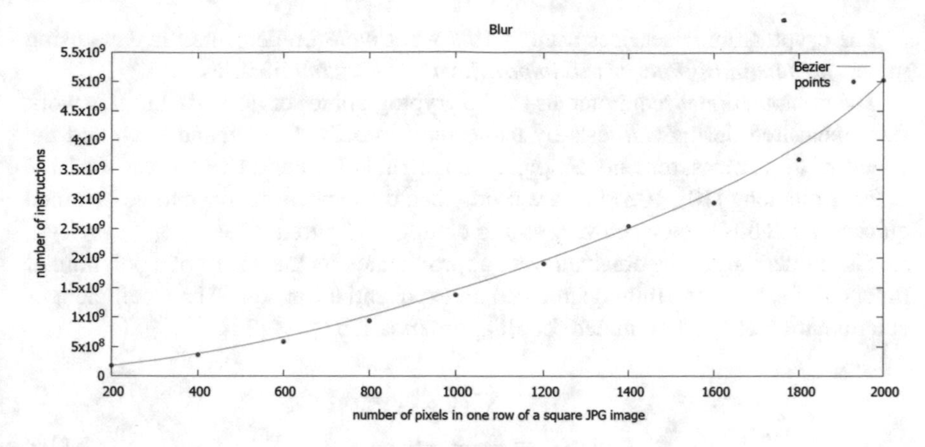

Fig. 2 Computational power required to blur an image

The workload is composed of image processing tasks that apply a blur filter to an image of a given size, such as 200 × 200, 2000 × 2000, etc. pixels, as illustrated in Fig. 2. The details about tasks features are presented in Table 2. In this experiment, a single batch composed of one task of each kind of those presented in Table 2 was used. This workload requires 19534 millions of instructions (MI) to be complete. Each task is modeled by using the $Cloudlet$ CloudSim class, while the $length$ parameter represents the computational requirements in MI, the *fileSize* parameter describes the size of the task inputs (*inputSize*) as shown in Eq. 3, and the *outputSize* parameter denotes the size of the result file.

Table 2 Tested task characteristics. Tasks were loaded as single batch. $outputSize_i = inputSize_i$ equals picture size in bytes

Task ID	picture dim. [pix. × pix.]	wl [MI]	*outputSize* [bytes]
1	200 × 200	180	136488 b
2	400 × 400	366	384992 b
3	600 × 600	582	661352 b
4	800 × 800	935	1001360 b
5	1000 × 1000	1381	1260784 b
6	1200 × 1200	1907	1764272 b
7	1400 × 1400	2541	2245408 b
8	1600 × 1600	2943	2747936 b
9	1800 × 1800	3672	3146304 b
10	2000 × 2000	5027	3421608 b
CloudSim CloudletID	CloudSim workloadvector	[MI]	[bytes]

The cryptographic services used in this work were implemented in Java using import *java.math.BigInteger* and *import java.util.Random* libraries.

The computational requirements of the cryptographic services used in this work were measured during run tests by using the Linux *Perf* command [28], and the results of these measurements are presented in Table 3. The value of n was set 1024 or 2048 bits long [10]. 1024 bits are used when the cypher text have to be obtained quicker, 2048 bits used when very strong coding is required.

The numerical results obtained were approximated by the means of a polynomial function of second and third degree and an exponential function. The Coefficient of determination R^2 was examined for all approximation types [29]:

$$R^2 = \frac{\sum_{t=1}^{n} (\hat{y}_t - \bar{y})^2}{\sum_{t=1}^{n} (y_t - \bar{y})^2} \tag{12}$$

where y_t is the empirical value, and $\hat{y}_t$ is the value obtained by using polynomial or exponential fitting. Finally, $\bar{y}$ is the mean value of the empiric values. The coefficient of determination, R^2, shows the quality of the approximation.

From among many scheduling methods, the Expected Time to Compute (ETC) matrix was used.

In order to estimate the time required to compute a particular task the scheduler needs the computing capacity (MIPS) for all VMs available, which can be denoted as the following vector:

$$CC = [cc_1, \dots, cc_m] \tag{13}$$

The tasks may be described by their computing requirements (MI) as a vector:

Table 3 Cryptographic coat for secret sharing, in MI

VMs nb.	Shamir split	Shamir-Gauss retrieve	Shamir-Lagrange retrieve
10	39	234	49
20	159	1137	191
30	380	2737	268
40	655	4519	273
50	1126	24141	393
60	1638	31947	433
70	2322	38576	524
80	2907	51931	681
90	5573	78794	929
100	9960	101155	1041

$$WL = [wl_1, \ldots, wl_n] \tag{14}$$

To compute the schedule solutions, the Expected Time to Compute (ETC) matrix may be used [30] for each VM i and task j as follows:

$$ETC = [wl_j / cc_i]_{i=1,2\ldots,m}^{j=1,2\ldots,n} \tag{15}$$

In order to take into account the security layer applied to tasks, the ETC matrix must be extended, and therefore the Security Biases Expected Time to Compute (SB_ETC) matrix [24] is used:

$$SBETC[j][i](SD, TL) = wl_j / cc_i + b(sd_j, w_j, tl_i, cc_i) \tag{16}$$

$$SBETC(SD, TL) = SB(SD, TL) + ETC. \tag{17}$$

The bias b is the time required to compute the RSA blinding procedure on a single picture. On the other hand, the bias B represents the rest of security operations. The RSA key generation is done by the user, and therefore it is omitted. The data transferring times were not taken into consideration in this simulation. The main scheduling objective is to minimize makespan:

$$C_{\max} = \min_{S \in Schedules} \left\{ \max_{j \in Tasks} C_j \right\}, \tag{18}$$

where C_j is the time when task j is finished. *Tasks* are all tasks submitted to the system and *Schedules* is the set of all possible schedules, as shown in Fig. 3.

5.2 Numerical Tests

5.2.1 Security Operations Scalability

The total computing effort required to share the Secret only depends on the number of participants, as shown in Table 3. Hence, the most efficient approach is to minimize the number of VMs by choosing those with a larger computational capacity, as shown in Fig. 4. Regarding this, the degree of the polynomial $dega(x) = t - 1$ presented in Eq. 8 is equal to the number of participants that hold the shares of the secret. To this aim, the relation between the number of participants $N \in Z : N \in [2, n]$, denoted by $N_{Sha,split}$, and the number of required instructions to compute the shares, denoted by $O_{Sha,split}$, was computed by the means of the Least Squares method [31], as described in Table 3.

As presented in Table 3, 39 MI are required to compute the Shamir secret spread process if ten worker VMs are used, while 159 MI are required if 20 worker VMs are used. These 39 MI last 0.054 s if the Dealer unit is deployed on m1.small instances,

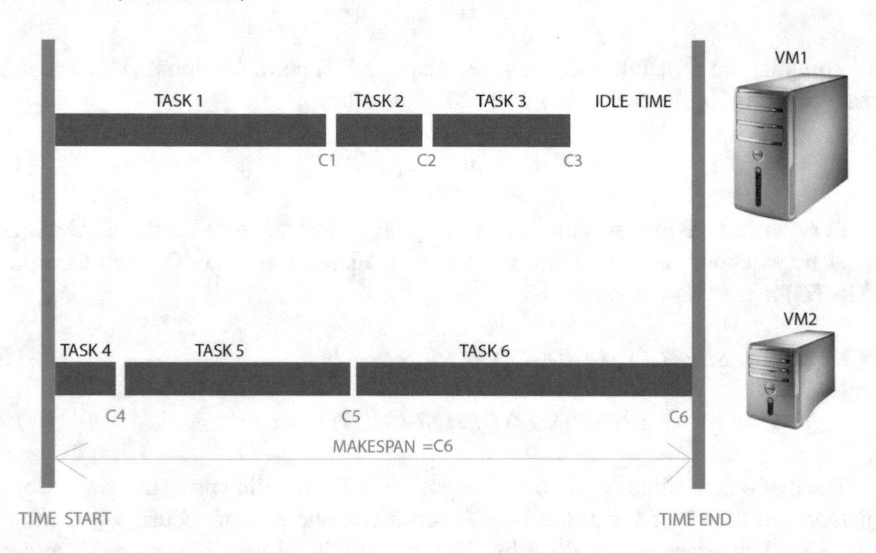

Fig. 3 Makespan computing process

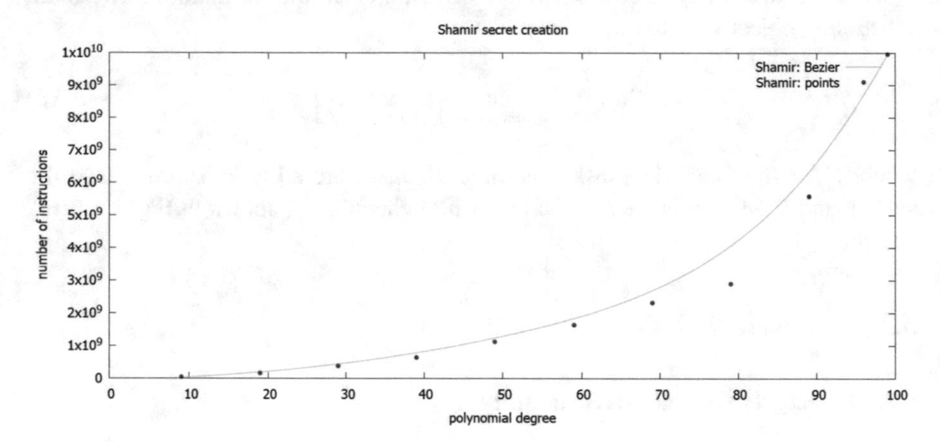

Fig. 4 Shamir secret creation computational effort in number of instructions

0.027 s if deployed on m1.medium instances, and 0.015 s if m3.large instances are used. The dependence between x, which represents the number of tasks, and s, which represents the size of the data files containing the image can be denoted as follows: $s = 11974.8 * x^2 + 253512x - 178297$.

The computational requirements of the RSA coding increase notably according to the task output size. These requirements, denoted by $N_{RSA,code}$ for the 1024-bit RSA and by $N_{RSA2048,code}$ for the 2048-bit RSA, were approximated by the means of polynomial and exponential functions, as shown in Table 4.

For the 2048-bit RSA, the best coefficient of determination was obtained for the third degree polynomial, and similar results are obtained for the 1024-bit RSA. The 2048-bit RSA is more demanding and these computational demands increase according to the output size faster than those of the 1024-bit RSA as illustrated in Fig. 5. As stated in Table 2, if 10 tasks are considered, the 1024-bit RSA requires 2853927 MI. This represents 4019 s if m1.small instances are used, while 1995 and 1150 s are needed to compute the encoding for m1.medium and m3.large instances, respectively.

The computational effort required to retrieve the Secret follows the same trend described for spreading the Secret and presented in Fig. 4.

The Secret retrieving process may be done by the means of two approaches: (a) Solving a set of linear equations by using the Gauss elimination method, as described in Eqs. 9–11, and (b) Using the Lagrange interpolation for a given point of the polynomial a. However, these methods present few differences, as described in Fig. 6 and Table 3.

The relation between the number of participants $N \in \mathbb{Z} : N \in [2, n]$, denoted by $N_{Sha,split}$, and the number of instructions required to compute the shares $O_{Sha,split}$ was computed by using the Least Squares method, [31], as shown in Table 3.

Regarding this, 234 MI were required when 10 VM were used, while 1137 MI were required for 20 VM. This means that the secret retrieving lasts 3.29 s if m1.small instances are used, while it lasts 0.16 and 0.09 s when m1.medium and m3.large are used respectively if 10 virtual machines are considered.

The computational effort required for decoding the result depends on the size of the picture that was blurred, as shown in Table 4. Regarding the bits used for RSA coding and decoding, it should be taken into account that the times required to compute 2048-bit RSA increases non-linearly as illustrated in Fig. 7. Thus, for the largest tasks, the 1024-bit RSA should be used, or the VM responsible for this computation should be scaled properly according to the results of this simulation.

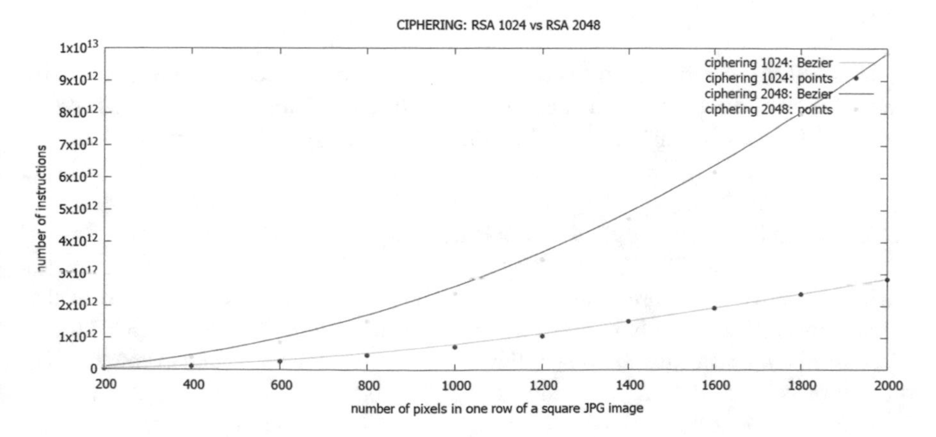

Fig. 5 Computational requirements for 1024-bit and 2048-bit RSA key

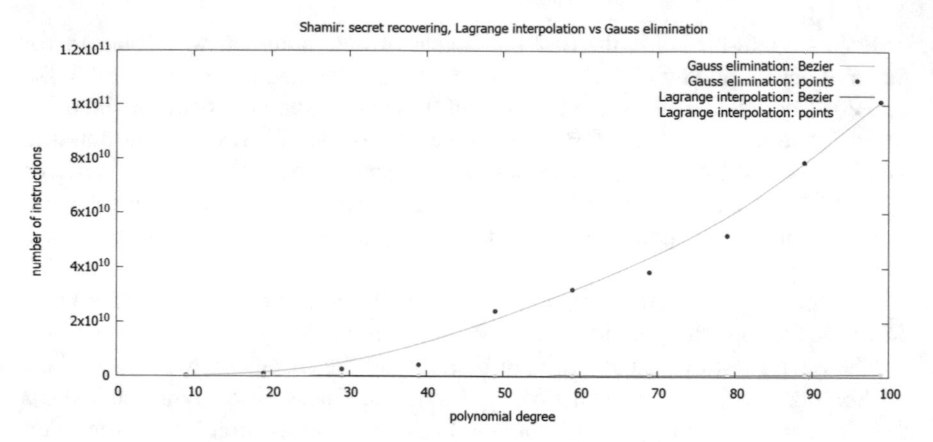

Fig. 6 Computational requirements for the two methods used for retrieving the Shamir secret

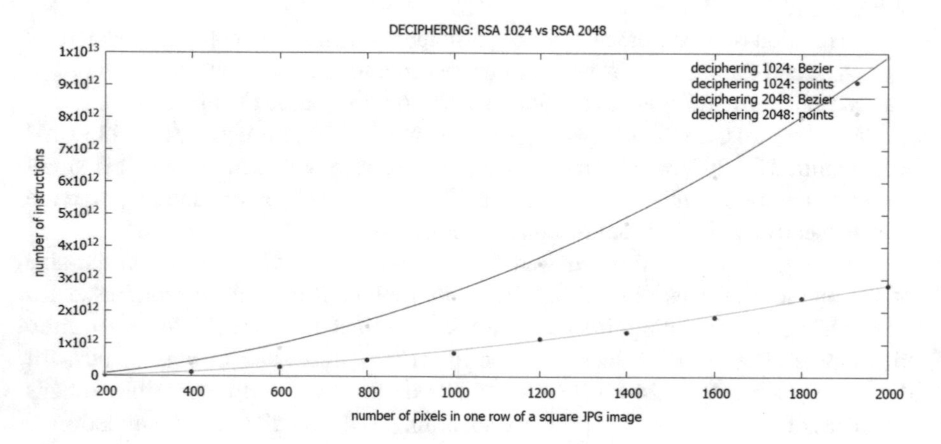

Fig. 7 Number of instructions for RSA 1024 versus RSA 2048 decoding

2814420 MI were required to compute this decoding process for the task ID=10 when the 1024-bit RSA is considered, as described in Table 2. This means 3963, 1995 and 1134 s when m1.small, m1.medium and m3.large are used respectively.

5.2.2 Security Layer Applied to the Batch of Tasks

Various security levels are considered regarding the security layer applied to the batch of tasks. These levels, denoted by a numerical value [0, 1, 2, 3] for both *TL* and *SD*, describe the strategies used as follows: (a) $TL = 0$ means that no cryptographic procedure is used, (b) $TL = 1$ denotes that the 1024-bit RSA coding is used, (c) $TL = 2$ means that both 1024-bit RSA coding and Secret sharing are used, and (d) $TL = 3$ that both 2048-bit RSA coding and Secret sharing are used.

For 10 worker nodes, the following results were obtained

- When $TL = 0$ is applied, the smallest task ID = 1 requires 180 MI, while the largest task ID = 5 requires 5027 MI.
- When $TL = 3$ is applied to the smallest task ID = 1, the secret spreading process adds 39 MI, while 95619 MI were added for the coding and blinding process, as described in Table 4. In addition, the secret retrieving process adds 49 MI, while the RSA decoding adds 93528 MI. Thus, The ratio between actual task operations and security operations for this small task is: $(39 + 95619 + 49 + 93528)/180 = 1051.3$
- When $TL = 3$ is applied to the largest task ID = 5, the secret spreading process adds 39 MI, while 9842010 MI were added for the coding and blinding process, as described in Table 4. In addition, the secret retrieving process adds 49 MI, while the RSA decoding adds 9884733 MI. Thus, for this large task, the ratio between actual task operations and security operations is: 3924.1 (see Table 4).

5.2.3 Security Impact on Makespan

In the proposed model, the makespan is only affected by the RSA blinding process, as stated in Eq. 18. When $TL = 0$ is applied, the maskespan may be computed by the means of the ETC matrix, as shown in Eq. 15. On the other hand, if all the machines offer $TL = 1$, 2 or 3, an additional cost has to be taken into account as shown in the following equation:

$$SB(SD, TL) = [b(sd_j, wl_j, 1, cc_i, inputSize_i, outputSize_i)]_{i=1,2...,m}^{j=1,2...,n} = \quad (19)$$

$$[\text{time of RSA blinding}]_{i=1,2...,m}^{j=1,2...,n}$$

Table 4 Cryptographic coat for RSA cypher, in MI

Task	Blind	1024 cod.	1024 decod.	2048 cod.	2048 decod.
1	2*0.1	29814	28434	95619	93528
2	2*0.3	114485	118751	383009	379847
3	2*0.6	263890	285172	861590	862114
4	2*1.0	469494	514528	1534304	1537535
5	2*1.2	734069	733254	2407671	2399740
6	2*1.7	1057640	1161299	3471082	3459379
7	2*2.2	1520914	1356978	4728474	4712922
8	2*2.7	1938632	1822616	6179446	6160087
9	2*3.1	2386142	2435027	7971389	8002257
10	2*3.4	2853927	2814420	9842010	9884733

Table 5 Curve approximation of number of operation for cryptographic procedures code, where x represents the number of tasks as shown in Table 2

Fitted curve	R^2
$Sha, split(N_{Sha,split}) = 1.44627 * 10^6 N^2_{Sha,split} - 9.22222 * 10^7 N_{Sha,split} + 1.48058 * 10^9$	0.607717
$O_{Sha,split}(N_{Sha,split}) = 5.44381 * 10^4 N^3_{Sha,split} + +4.64971 * 10^2 N^2_{Sha,split} + 1.04114 * 10^7 N_{Sha,split} - 1.42388 * 10^9$	0.988185
$N_{RSA,code} = 1.73458 * 10^9 x^3 - 1.229 * 10^{10} x^2 + 2.02758 * 10^{11} x - 2.14152 * 10^{11}$	0.987939
$N_{RSA,code} = 1.85489 * 10^{11} e^{0.279988x}$	0.988229
$N_{RSA2048,code} = 1.01126 * 10^9 x^3 + 8.55498 * 10^{10} x^2 + 3.15861 * 10^{10} x - 2.58364 * 10^{10}$	0.999958
$N_{RSA2048,code} = 5.527 * 10^{11} e^{0.292838x}$	0.992466
$O_{Sha,retr}(N) = 1.82627 * 10^3 N^3 - 2.03002 * 10^5 N^2 + 1.63507 * 10^7 N - 8.41731 * 10^7$	0.989524
$N_{RSA,decode}(x) = -9.31435 * 10^7 x^3 + 2.75375 * 10^{10} x^2 + 2.1635 * 10^{10} x - 2.39607 * 10^{10}$	0.996551
$N_{RSA,decode}(x) = 1.84664 * 10^{11} e^{0.278962x}$	0.989757
$N_{RSA2048,decode} = 1.5414110^9 x^3 + 7.8444510^1 0x^2 + 5.6787610^1 0x - 4.9506610^1 0$	0.999923
$N_{RSA2048,decode}(x) = 5.4689310^1 1e^{(0.294253x)}$	0.992726

It should be kept in mind that using a light-weight secret sharing strategy has only 5% of negative impact in terms of makespan, as presented in Table 7.

6 Conclusions and Future Development

In this work, various security services based on result encryption are presented. The described approach, which stores encrypted tasks results by the means of a blinded RSA algorithm, increase notably the level of security of the Cloud Computing systems.

Regarding this, the Cloud Storage Unit is the responsible for receiving the blinded results and encrypting them by using the proper cryptographic keys. The final user or further processing units may decrypt this result on demand by the means of the Shamir secret approach instead of a hash function.

The following conclusions can be stated in order to summarize this work:

- The proposed cryptographic layer may be divided in: (a) Operations having an impact on the scheduling process, and (b) Operations not impacting the scheduling process.

Table 6 The computational time savings using different type of instances

VM type	TL	Task	Time [s]	% Time saved
m1.small	0	1	0.253	0%
m1.med	0	1	0.125	100 – 12.5/0.253 = 50.59%
m3.large	0	1	0.072	100 – 7.20/0.253 = 71.54%
m1.small	0	5	1.945	0%
m1.med	0	5	0.965	100 – 96.5/1.945 = 50.39%
m3.large	0	5	0.556	100 – 55.6/1.945 = 71.41%
m1.small	0	10	7.080	0%
m1.med	0	10	3.515	100 – 351.5/7.080 = 50.35%
m3.large	0	10	2.027	100 – 202.7/7.080 = 71.37%
m1.small	1	1	82.29	0%
m1.med	1	1	40.85	100 – 4085/82.29 = 50.36%
m3.large	1	1	23.55	100 – 2355/82.29 = 71.38%
m1.small	1	5	2068.59	0%
m1.med	1	5	1027.06	100 – 102706/2068.59 = 50.35%
m3.large	1	5	592.21	100 – 59221/2068.59 = 71.37%
m1.small	1	10	7990.66	0%
m1.med	1	10	3967.39	100 – 396739/7990.66 = 50.35%
m3.large	1	10	2287.65	100 – 228765/7990.66 = 71.37%
m1.small	2	1	82.41	0%
m1.med	2	1	40.92	100 – 4092/82.41 = 50.35%
m3.large	2	1	23.59	100 – 2359/82.41 = 71.37%
m1.small	2	5	2086.22	0%
m1.med	2	5	1035.81	100 – 103581/2086.22 = 50.35%
m3.large	2	5	597.26	100 – 59726/2086.22 = 71.37%
m1.small	2	10	7990.79	0%
m1.med	2	10	3967.45	100 – 396745/7990.79 = 50.35%
m3.large	2	10	2287.68	100 – 228768/7990.79 = 71.37%
m1.small	3	1	266.78	0%
m1.med	3	1	132.45	100 – 13245/266.78 = 50.35%
m3.large	3	1	76.37	100 – 7637/266.78 = 71.37%
m1.small	3	5	37192	0%
m1.med	3	5	18466	100 – 1846600/37192 = 50.35%
m3.large	3	5	10647	100 – 1064700/37192 = 71.37%
m1.small	3	10	27791	0%
m1.med	3	10	13798	100 – 1379800/27791 = 50.35%
m3.large	3	10	7956	100 – 795600/27791 = 71.37%

Table 7 The impact of proper scheduling and Shamir secret sharing procedure into makespan, test for 5 VMs: all instances types included, batch of tasks built from tasks ID = 1, ID = 2, …, ID = 10

VM number	TL	WL of batch [MI]	Initial makespan [s]	Final makespan [s]
5	0	19534	11.99	2.13
5	1, 2, 3	19534 + 33	11.99	2.42

- These operations may be chosen according to the kind of task considered. However, it should be kept in mind that the number of worker nodes may have a large impact on the performance of the cryptographic operations. Regarding this, various strategies may be adopted in order to keep the overall performance stable, such as: (a) Tuning the number of instances used, (b) Tuning the type of instances used, and (c) Tuning the parameters used in the cryptographic operations.

Finally, the described model is valuable to improve the makespan of the tasks executed in Cloud Computing systems. As future work, a further step towards the automation of these security processes applied to Cloud Computing systems should be explored, as described in [32].

Acknowledgements This chapter is based upon work from COST Action IC1406 High-Performance Modelling and Simulation for Big Data Applications (cHiPSet), supported by COST (European Cooperation in Science and Technology).

References

1. Amazon Web Services: https://aws.amazon.com
2. Google Cloud: https://cloud.google.com
3. Microsoft Cloud: http://www.microsoft.com/enterprise/microsoftcloud
4. Adobe Creative Cloud: http://www.adobe.com/pl/creativecloud.html
5. Cloud Controls Matrix Version 3.0.1, Cloud Security Alliance: https://cloudsecurityalliance.org/group/cloud-controls-matrix/
6. Directive of the European Parliament and of the Council: On the protection of individuals with regard to the processing of personal data and on the free movement of such data. http://eur-lex.europa.eu/legal-content/EN/TXT/?uri=CELEX:31995L0046 (1995)
7. Stinson, D.R.: Cryptography: Theory and Practice. CRC Press (2005)
8. Mell, P.M., Grance, T.: The NIST definition of cloud computing. SP 800-145. Technical Report (2011)
9. Jakbik, A., Grzonka, D., Koodziej, J.: Security supportive energy aware scheduling and scaling for cloud environments. pp. 583–590 (2017). https://www.scopus.com/inward/record.uri?eid=2-s2.0-85021827530&partnerID=40&md5=30d087573993bf732184bee293687bce. Cited by 0
10. NIST Cloud Computing Standards Roadmap. SP 500-291, Version 2: Technical Report. https://www.nist.gov/sites/default/files/documents/itl/cloud/NIST_SP-500-291_Version-2_2013_June18_FINAL.pdf (2013)

11. Gilly, K., Juiz, C., Thomas, N., Puigjaner, R.: Adaptive admission control algorithm in a QoS-aware web system. Inf. Sci. **199**, 58–77 (2012). https://doi.org/10.1016/j.ins.2012.02.018, https://doi.org/10.1016/j.ins.2012.02.018
12. Gupta, H., Dastjerdi, A.V., Ghosh, S.K., Buyya, R.: iFogSim: a toolkit for modeling and simulation of resource management techniques in internet of things, edge and fog computing environments. CoRR **abs/1606.02007** (2016). http://arxiv.org/abs/1606.02007
13. Suchacka, G., Borzemski, L.: Web Server Support for e-Customer Loyalty Through QoS Differentiation, pp. 89–107. Springer, Berlin Heidelberg (2013). https://doi.org/10.1007/978-3-642-53878-0_5, https://doi.org/10.1007/978-3-642-53878-0_5
14. Jakóbik, A.: Big Data Security, pp. 241–261. Springer, Cham (2016). https://doi.org/10.1007/978-3-319-44881-7_12
15. Jakóbik, A., Grzonka, D.: Energy efficient scheduling methods for computational grids and clouds. J. Telecommun. Inf. Technol. (2017)
16. Secure Hash Standard: Technical Report. https://doi.org/10.6028/NIST.FIPS.180-4 (2015)
17. Shamir, A.: How to share a secret. Commun. ACM **22**(11), 612–613 (1979). https://doi.org/10.1145/359168.359176
18. Schubert, G.R.: Algorithm 210: Lagrangian interpolation. Commun. ACM **6**(10), 616 (1963). https://doi.org/10.1145/367651.367665
19. Gauss, E.J.: A comparison of machine organizations by their performance of the iterative solution of linear equations. J. ACM **6**(4), 476–485 (1959). https://doi.org/10.1145/320998.321001, https://doi.org/10.1145/320998.321001
20. CloudSim: https://github.com/Cloudslab/cloudsim/
21. Buyya, R., Ranjan, R., Calheiros, R.N.: Modeling and simulation of scalable cloud computing environments and the cloudsim toolkit: challenges and opportunities. In: 2009 International Conference on High Performance Computing Simulation, Leipzig, 2009, pp. 1–11. https://doi.org/10.1109/HPCSIM.2009.5192685
22. Amazon Cloud EC2 Instance Types Tests: https://www.ec2instances.info/
23. Amazon EC2 Instance Types Tests: http://www.cloudlook.com/amazon-ec2-m1-medium-instance
24. Jakóbik, A., Grzonka, D., Kołodziej, J., Gonzalez-Velez, H.: Towards secure non-deterministic meta-scheduling for clouds. In: Proceedings of 30th European Conference on Modelling and Simulation, ECMS 2016, Regensburg, Germany, May 31–June 03, 2016, pp. 596–602. https://doi.org/10.7148/2016-0596
25. Jakbik, A., Grzonka, D., Palmieri, F.: Non-deterministic security driven meta scheduler for distributed cloud organizations. Simulation Modell. Pract. Theory **76**, 67–81 (2017). ISSN 1569-190X. https://doi.org/10.1016/j.simpat.2016.10.011
26. Grzonka, D., Jakbik, A., Kołodziej, J., Pllana, S.: Using a multi-agent system and artificial intelligence for monitoring and improving the cloud performance and security. Future Gener. Comput. Syst. (2017). ISSN 0167-739X. https://doi.org/10.1016/j.future.2017.05.046
27. Amazon EC2 Instances: https://aws.amazon.com/ec2/instance-types/
28. Linux Perf Command: https://perf.wiki.kernel.org/index.php/Main_Page
29. Knill, O.: Probability and Stochastic Processes with Applications. Overseas Press (1994)
30. Koodziej, J.: Evolutionary Hierarchical Multi-criteria Metaheuristics for Scheduling in Large-Scale Grid Systems. Springer (2012)
31. Bayen, A.M., and Siauw, T.: Chapter 12–Linear Algebra and Systems of Linear Equations, In An Introduction to MATLAB® Programming and Numerical Methods for Engineers, Academic Press, Boston, 2015, pp.177–200. ISBN 9780124202283. https://doi.org/10.1016/B978-0-12-420228-3.00012-9
32. Jakbik A., Wilczynski, A.: Using polymatrix extensive stackelberg games in security aware resource allocation and task scheduling in computational clouds. J. Telecommun. Inf. Technol. (2017)

Printed in the United States
By Bookmasters